Contemporary Inquiry in Geography

Research Lecture Series-1

CONTEMPORARY INQUIRY IN GEOGRAPHY

(RESEARCH LECTURE SERIES - I)

Contemporary Inquiry in Geography

Research Lecture Series-1

Edited by
Rash Bihari Prasad Singh
Vishwa Nath Prasad Sinha
Poornima Shekhar Singh

RAJESH PUBLICATIONS
NEW DELHI - 110 002

Contemporary Inquiry in Geography

(Research Lecture Series-1)

Published by :
Sanjay Gupta
RAJESH PUBLICATIONS
1, Ansari Road, Darya Ganj, New Delhi - 110002
Ph.: 011-23274550, 43084725
Website : www.rajeshpublications.com
Email : rajeshpublications@gmail.com

First Published in India 2021

ISBN : 978-93-91139-10-0

Price : Rs. 500.00

Printed in India by :
Aryan Digital Press
Delhi - 110002

From Editorial Desk

No academic is an island. The Centre for Geographical Studies, A.K.U, Campus, Patna started to build an academic environment with excellence in the midst of Covid-19 pandemic. The Centre took the corona protocols as like, social distancing and isolation/quarantine as a stimulating force and resolved to make use of this dark face of social spectrum as an opportunity to establish academic credential and moved forward from academic island to create academic mosaic by creating a diverse intellectual environment where large number of established scholars, teachers, students and social activities were encouraged to remain academically active and sound.

The Centre for Geographical Studies, therefore, started a web-based lecture series on contemporary social, economic and environmental challenges being faced by the mankind. In fact, it was a great take-off by a newly created academic platform. Altogether 17 web- based lecture have already been delivered. This web-based lecture series was widely circulated and greatly appreciated. The Director of the Centre took another major resolution to publish all the lectures as printed document, so that the new generation is benefited. At this moment, we, editors got an opportunity to arrange, edit and systematise the papers presented by the scholars. Taking pen in the hand to pen the scholary academic documents is extremely painful. But it has been academic custom to make the contents presentable in the printed format without removing / editing/replacing the basic content objective and findings of the writers. I along with all members of the editorial board have been careful in putting the contents in the book's format. The Centre as well as editorial board shall not be responsible for any part of the content, in case it is contested. All themes, contents and ideas are personal views of respective scholars. They shall be liable for any kind of discrepancy. We hope that the publication of this book will be appreciated as the most nobel way of making use of time during the corona period.

Reader's suggestions will always be welcome. Reasonable suggestions will definitely be incorporated in the next edition of the book. At the last but not the least, I along with all members of editorial board express sincere thanks to the scholars who contributed through web-based lecture and also to the Director and other fellow members of the Centre for providing all kinds of support whenever and wherever needed. We are also thankful to Rajesh Publications, New Delhi for agreeing to publish the book in a very short notice.

Raksha Bandhan
(Shrawanipurnima)

The 22th August, 2021 (Sunday)
Place: Patna

Rash Bihari Prasad Singh
Editor-in-chief

Editorial Board Members
Prof. V.N.P. Sinha
Prof. Poornima Shekhar Singh

From the Pen of Director

The Centre for geographical studies, Patna is an unique initiative of the Government of Bihar, Patna. This is first of its kind in the country. The Government and academia realize importance of geographical methodology and analysis in development sectors. Geographical analysis can be of great help in deciding the allocation of resources in resource poor society. This will immensely help in balanced regional growth. For example, if development of educational infrastructure is decided by doing spatial analysis, the resources will be better utilized. Regional and human disparities can also be better addressed by applying appropriate geographical methodologies.

The Centre for Geographical Studies is affiliated to Aryabhatta Knowledge University, Bihar for giving degrees to research scholars in the field of Geography. The Centre has been forging academic collaboration and partnership with similar organisations/institutions to develop integrated methodology in order to strengthen geographical research in the fields of development and welfare. The Centre has also started several academic and research activities which may be helpful to various department of state government at the planning, development and implementation stages. The Centre is also engaged in developing geospatial information system that can be used by various organisations and will be offering course to train the geographers and others in the field of this technology. Thus, it intends to build a pool of skilled personnels.

The Centre is in its nascent stage and during pandemic it has successfully organized Web-based Lecture Series on various important issues involving national and international importance. Some of the important speakers were from Jawaharlal Nehru University, New Delhi; International Institute for Population Studies, Mumbai; Centre for Environment Planning & Technology, Ahmedabad; Aligarh Muslim University, Aligarh; North-Eastern Hill University, Shillong; Patna University, Patna besides one foreign speaker. The Centre encourages multidisciplinary approach and involves not only geographers but also planners, geologists and social scientists in its activities.

The Centre recognizes importance of quality publication in the field of Geographical Research. Various publications are in the pipe line and will be available to public at large in recent future. The Centre also offers Ph.D. programme to qualified students and right now three Ph.D. students are engaged working on important issues. Finally, the Centre aims to become a pioneer research organization in the field of geography in the country.

Moreover, we are publishing this edited book to consolidate our academic achievements even in the pandemic situation. We are also hosting XVI DGSI International Geography Online Conference from 10th September to 12th September, 2021. I am happy to inform the academia that this edited book will be released on this occasion.

Raksha Bandhan
(Shrawani Purnima)
The 22nd August, 2021 (Sunday)
Place: Patna

Prof. Poornima Shekhar Singh
Director
Centre for Geographical Studies
AKU Campus, Patna

CENTRE FOR GEOGRAPHICAL STUDIES
ARYABHATTA KNOWLEDGE UNIVERSITY CAMPUS,
MITHAPUR, PATNA-800001
E-MAIL: cgs.aku@gmail.com

List of Contributors

1. **Prof. Rash Bihari Prasad Singh**
Ex. Vice-Chancellor, Nalanda Open University and Patna University, Patna, Bihar, India

2. **Dr. Anil Kumar Roy, FeSRA**
Associate Professor, Faculty of Planning, CEPT University, K.L. Campus, University Road, Navrangpur, Ahmedabad, Gujarat - 380009

3. **Prof. Abha Laxmi Singh**
Professor (Retd.) of Geography, Dept. of Geography, Aligarh Muslim University, Aligarh, 202002

4. **Prof. Atul Aditya Pandey**
Head of the Department of Geology, Patna University, Patna, Bihar, India

   Prof. Ramesh Shukla
   Professor of Geology (Retd.), Department of Geology, Patna University, Patna, Bihar, India

5. **Akhouri Bishwapriya**
Director, Geological Survey of India, SU: Bihar, Patna

   Priya Upadhyay
   Geologist, Geological Survey of India, SU: Bihar, Patna

   Abhishek Kumar Chaurasia
   Assistant Geologist, Geological Survey of India, SU: Bihar, Patna

   Shikha Dagar
   Geophysicist, Geological Survey of India, SU: Bihar, Patna

   G.S. Gokul
   Geophysicist, Geological Survey of India, Eastern Region, Kolkata

   Sandeep Kumar
   Geophysicist, Geological Survey of India, Eastern Region, Kolkata

6. **Dr. Tapas Pal**
Assistant Professor, Dept. of Geography, Raiganj University, West Bengal, India

7. **Prof. R.B. Bhagat**
Professor and Head, Department of Migration and Urban Studies, International Institute for Population Sciences, Mumbai-400088

8. **Dr. Arun Kumar**
Senior Scientist, Mahavir Cancer Sansthan and Research Centre, Patna, Bihar, India

Prof. Ashok Kumar Ghosh
Head of the Department, Research Centre, Mahavir Cancer Sansthan and Research Centre, Patna, Bihar, India

9. **Dr. Saleha Jamal**
Associate Professor, Dept. of Geography, Aligarh Muslim University, Aligarh-202002

10. **Prof. Sarfaraz Alam**
Department of Geography, Institute of Science, Banaras Hindu University, Varanasi-221005, Uttar Pradesh, India

11. **Dr. Ashish Ranjan**
National Convenor of National Alliance of People's Movements and Security, Jan Jagran Shakti Sangathan

12. **Rajiv R. Thakur (Ph.D.)**
Associate Professor of Geography, Missouri State University, West Plains, Missouri, U.S.A.

13. **Dr. Habibullah Ansari**
Associate Professor & Head, Division of Social Psychology, A.N. Sinha Institute of Social Sciences, Patna

Contents

1

Importance of Geographical Research in the Sustainable Development of India

Rash Bihari Prasad Singh
singh.rbp@gmail.com

Introduction

Research is an academic exercise to attain some kinds of new ideas having fruitful relevance for all the development, welfare, excellence and quality improvement of the mankind and its milieu, Geographical research around the world has also been found useful not only for academia but also for spatial planning in order to achieve the development and welfare goals with high level of satisfaction within the time frame.

Since the Agenda-21 declaration (1992), there has been paradigm shift in the research priority and its Milieu in most of disciplines due to changes in the basic objective of development. This change is from socio-economic development to sustainable development. The United Nations General Assembly in its 70th session held on 25th September, 2015 adopted the document titled Transforming our world: The 2030 Agenda for Sustainable Development, also known as AGENDA 2030 with effect from January 1, 2016. Since then, there has been global efforts to achieve the targets from local, regional and national to global level. Sustainable development strategy is addressed to make the use of available resources and heritage of the nature / environment to such scale/level that the endowed resources also remain conserved for the future generation.

The implementation of this paradigm shift in the planning and development strategies is not an easier task. It requires rationalisation of the use of resources, appraisal of environmental elements, estimation of next generation needs, conservation planning for lithosphere, hydrosphere, biosphere and atmosphere. Long term projection for consumption, conservative and continuum of nature and peoples are needed. It is this situation where geographers have a role. Geographers principally study man- environment relationship in the perspectives of spatial pattern, process and organisation. My objective of presenting this lecture is to explore the areas where geographers can contribute significantly in the sustainable planning and development of defined spaces from local to national level.

Importance of Geographical Research in planning and development:

Classroom teaching, research, experiments and field work are four major pedagogical components of any discipline. Since the rise of post - modern geography, there has also been a tendency of interdisciplinary inquiries. As like other disciplines, post-modern geography has also received drastic changes in the methodology, nature and dimension of research themes.

Chronically, there are four important phases of geographical research. They are:

1. Descriptive research
2. Analytical research
3. Synthesised research
4. Pragmatic research

Descriptive research period was pre- second world war period. Findings were based on observation, assessments and concerned descriptive literature. In fact, findings were based on personal observation and assessment and that caused personal bias in research findings. Consequently, findings on the identical topic were found varying and due to that pre- industrial revolution Geographical research failed to be relevant in the social, economic and cultural context of the mankind.

After 2nd world war, two important work brought significant change in the nature of geographical research. They were work of Schaefer (1953) on 'Philosophy and Method of Geography' and Hartshorne's emphasis on spatial analysis. Schaefer's (1959) work became the basis of quantitative revolution and research investigation based on quantitative tools started to provide analytical explanation of facts. Eratosthenes's (420BC) geographical philosophy was the study of earth and its related phenomena. In fact, theme of geographical study continues to remain the same by and large but approaches, tools of explanation and nature of inquiry received paradigm shift after 2nd world war. Hartshorne (1959) emphasised over the study of space in the context of spatial pattern, process and, organisation. Schaefer philosophy and method became the basis of positivism in geographical inquiries. In geography, BJL Berry's emphasis on quantitative tools further strengthen the ideas of positivism in geography. It was the time of positivism when some Indian geographers as like Dr. Praksha Rao, L.S. Bhatt. Moonis Raza, R.P. Misra and several others started to take interest in the planning and development inquires in the geographical perspective. Their continuous writings created an environment for India's area based planning. However, it was not in a position to have assertive impact over the decision taking process.

Post - Positivism period Research in Geography:

The work of David Harvey (1969) was the divide line in the use of tools approaches, methodology, mapping, space explanation and to some extent changes in the themes of explanation. The basic objective behind all these changes were to make geography anthropocentric and academically sound to address the issues of welfare, planning, Socio-economic development and optimal use of space where environmental conditions are conserved at the best possible level. Hagget (1972)'s book "Geography: A Modern Synthesis" and work of Ley and Samuels (1978) on pragmatic explanation in geography significantly improved the acceptability of geography as a subject having quality research for space analysis and regional planning. Furthermore, The work of Buttimer (1970) on humanistic approach, welfare and well-being approach of Smith (1977) and radical approach of Peet (1977) contributed in the improvement of research quality in human geography. They also

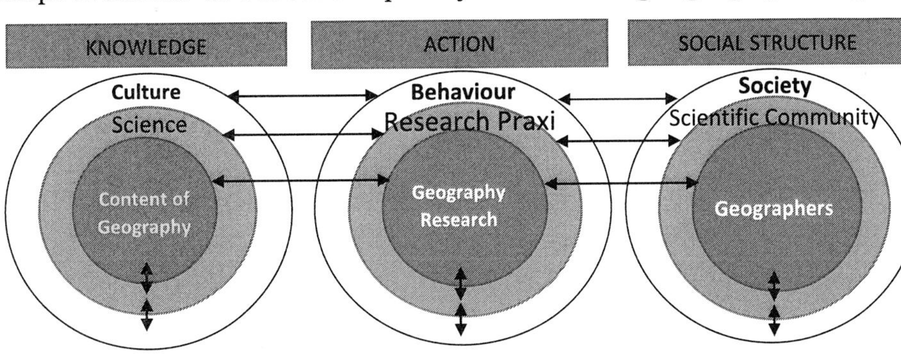

GRANO MODEL ON GEOGRAPHICAL RESEARCH (1981)

contributed in prompting the multidisciplinary research for spatial / regional planning. These changes brought distinctiveness and that was further credited due to high quality data generation and management techniques, multi-layered need-based mapping (GIS) and micro area analysis in terms of multi-layered space management. The given model (Grano, 1981) shows how the present-day geographical research has become indispensable for the sustainable development of both space and society.

Geographical Research and Sustainable Development Approach in India

The concept of sustainable development emerged from the Word Commission Report on Environment and Development (1987). It was globally accepted as eco-friendly development model in 1992 under Rio Declaration, Agenda-21. It was further strengthen by Johannesburg declaration in 2002. Since then, the world community understands that the sustainable development goals shall only be achieved by nature-friendly development approach. This kind of development will meet the need of the present generation but also the needs of future generation. There are 17 development goals and the NITI Aayog in India acts as the monitoring and performance assessment agency. All 17 goals are to be implements by 2030. These are:

1. Make India free from Poverty
2. Zero level of Hunger
3. Good health and well-being of the citizens
4. Quality Education
5. Gender Equality
6. Clean water and Sanitation
7. Affordable and clean energy
8. Decent work and Economic growth
9. Industry, innovation and infrastructure using appropriate and sustainable technology
10. Reduction of Inequalities
11. Sustainable Cities and Communities
12. Sustainable Consumption and production

13. Climate action
14. Life below water
15. Life on land
16. Peace, justice and strong Institution
17. Partnership for the goals.

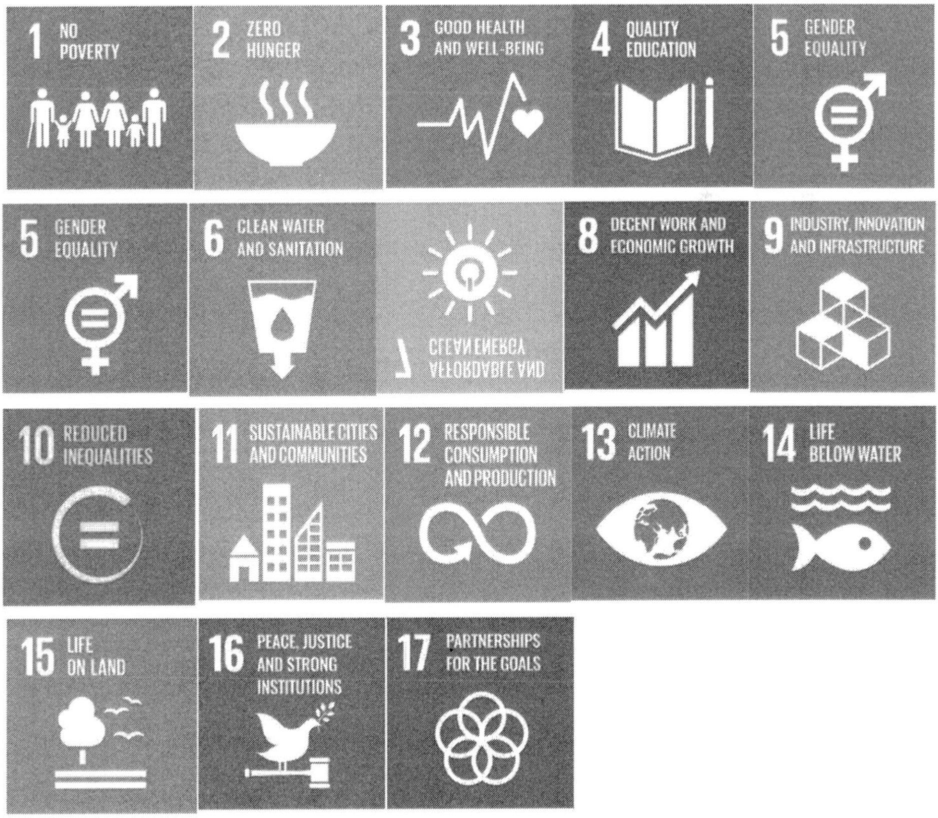

Symbolic Presentation of Sustainable Development Goals (2016-2030)

Approved and Adopted by UNO General Body

These 17 goals can not be sustainably achieved without giving due space to geographical research in the implementation strategy. Spatial pattern of poverty, hunger, well-being etc. are to be identified. Areas of highly developed, moderately developed and under developed are also to be recognized. Procedural errors are to be rectified so that the underdeveloped areas get new procedural and organizational support in achieving the targets. NITI Aayog has already started to involve geographers in identifying the causes and consequences of climate change and well-being planning and several other goals also need to be geographically synthesized for sustainable development strategy. Remote sensing information and GIS maps are being used to explore the areas of vulnerability and low sustainability. Ministry of Climate Change and Jalshakti have been created to achieve goals under the heads of 13 and 6 respectively. Make in India, Digital India, Aayushman Bharat, Smart City Programmes and Swachh Bharat Abhiyan are also launched to meet the sustainable development targets.

In most of cases, India's achievement level has been well below the targets. Here geographers are in a capacity to help the development agencies due to the better understanding on "Who" gets "What", "Where" and "How". Who is the target group population as like population living below poverty line? What covers the kind and degree of demand? Here is the key of sustainability, Gandhi has rightly said, there is no dearth of resources for your need but there is a dearth to meet your greed. This philosophy of Mahatma Gandhi is based on the principle of "Live and let them live." "Where" refers to spatial pattern as like the location of below poverty line population can lie spatially identified. "How" is related to spatio-temporal Process.

Rapid rise and voluminous size of population, massive use of natural resources massive deforestation, and some of the adverse facts continue to create alarming situation for the survival of the mankind. Sustainable Development policy exhibits a silver lining but the support of society, local community leaders, scientific community, political leadership and global partnership for parity, peace and prospering will be essential to implement the 17 goals of sustainable development. In this context geographers academic activities in the form of research data collection, data management and GIS mapping will play significant role. Once Sustainable development goals are successfully implemented, the alarming of threats of survival is expected to disappear and the planet will again be capable to support our needs not only today but for generations to come.

References:

Grano, O (1981): External Influence and Internal change in the Development of Geography in D. R. Stoddarh (ed) "Geography, Ideology and Social Concern", Oxford, Blackwell, 17-36

Haggett, P (1972): Geography: A Modern Synthesis, Harpa and Row,

New York.

Hartshorne, R (1959): Perspective on the Nature of Geography, Rand Mc Nally, Chicago

Harvey, D (1969): Explanation in Geography Edward Arnold, London

Harvey, M.E & Holly, B. P (2002): Themes in Geographical Thought, Rawat Publications, New Delhi

Johnston, R and Sidaway, J.D (2017, Reprint): Geography and Geographers, Routledge, New York

Kumar, Sanjay (2020) ed: Population, Environment, Empire Books International, New Delhi

Ley, D and Samuels, M.S.(1978): Introduction : Contexts of Modern Humanism in Geography in D. Ley and M.S. Samuels (eds) 'Humanistic Geography : Prospects and Problems, Maaroufa Press, Chicago, 1-18

Ministry of Finance, Govt. of India (2019); "India follows a holistic approach towards its 2020 sustainable Development Goals (SDGS).

Peet, R (1977): Radical Geography Maaroufa Press, Chicago, 6-30

Schaefer, F.K. (1953): Exceptionalism in Geography: A Methodological Examination, Annals of the Association of American geographers, 43, 226-49

Smith, D.M (1977) : Human Geography : A Welfare Approach, Edward Arnold, London

Tiwary, A.N. (2016): Development Studies in Geography: Concepts and Approaches, Asian Journal of Multi-Disciplinary Studies, Vil IV, PP 249-253

World Commission on Environment and Development/Brundtland commission (1987): Our Common Future, report Submitted to UNO, New York

2

Issues and Challenges of Urbanization in India

Anil Kumar Roy
anil.roy@cept.ac.in

Introduction

Urbanisation in India brings both challenges and opportunities at the same time. While it provides huge opportunities for labour markets in urban areas, the urban areas also become places of production of goods and services which are non-primary and their distributions and marketing. The urban areas in the process become the service centres of their respective hinterlands. There are various processes through which a settlement transforms into an urban area. They are i) natural increase of the population of a settlement beyond a limit of 5000 population; ii) migration of population from rural to urban areas, iii) change of the boundaries of urban settlement and iv) notifying a settlement into the urban area by the respective state governments. While these transformations are taking place, they bring both challenges and opportunities together. This chapter presents a brief account of how urban areas are defined in India and their process of urban transformation. It also presents a long term trend of urbanization. The chapter also deals with the issues and various challenges of urbanization in India. It concludes that urbanization in India is still slow, however, it is likely to pick up in near future compounding the issues and challenges that the urban settlements are facing.

Defining urban in India:

Urbanisation is the process through which rural settlements get transformed into urban, meaning that the former is predominantly an agrarian economy and later is not. On the contrary, urban settlements are defined in India based on statutory and non-statutory criteria. The statutory criteria are the legal status given by the respective state government to any settlement that may or may not fulfil non-statutory criteria i.e., given by the Registrar of Census of Population, Government of India. The entire urban settlements are categorized into **Statutory** and **Non-Statutory** (Census) towns in India. They are defined as the following.

All places with a municipality, corporation, cantonment board or notified town area committee, etc. are known as **Statutory Towns.**

All other places which satisfy the following criteria:

1. A minimum population of 5,000;
2. At least 75 per cent of the male working population engaged in non-agricultural pursuits; and
3. Population density of at least 400 persons per sq. km.

They are also known as **Census Towns (Non-statutory).**

Trends of urbanization (1901-2011)

The urban population of India includes the population living in these two categories of urban settlements. Hence, the level of urbanization i.e., the percentage of the population living in urban areas in India is 31.16 per cent as per the last census 2011, however, it is projected to be around 33 per cent in the next census in 2021. This shows that there has been an increasing trend in the share of the urban population in India in the previous decade. The long-term trends have also been increasing since the beginning of the last century. At the time of the beginning of the 20th Century India's urban population was only about 11 per cent, however, the share has increased to 17 per cent after the Independence in the census year 1951.

Table 1
Urban Populations, Share and Growth Rate during 1901-2011

Year	Total Population	Urban Population	Per Cent Urban	Decadal Growth Rate
1901	238396327	25851873	10.84	
1911	252093390	25941633	10.29	0.35%
1921	251321213	28086167	11.18	8.27%
1931	278977238	33455989	11.99	19.12%
1941	318660580	44153297	13.86	31.97%

1951	361088090	62443709	17.29	41.42%
1961	439234771	78936603	17.97	26.41%
1971	598159652	109113977	18.24	38.23%
1981	683329097	159462547	23.33	46.14%
1991	844324222	217177625	25.72	36.19%
2001	1028737436	286119689	27.81	31.74%
2011	1210193422	377105760	31.16	31.8%

Sources: Various Censuses 1901 to 2011

The share of the urban population could reach 28 per cent at the beginning of the 21st Century. The last census (2011) reported 377 million urban population in India with a share of 31 per cent of the total population, it is expected that about one-third of the total population would live in urban areas by 2021. The share of the urban population in India may look less compared to other developing countries such as Brazil, South Africa, Indonesia, Malaysia and China, however, the absolute number of populations living in urban India is next to China (Aijaz, 2021). India's total urban population is a little larger than the total population of the USA.

According to mincrotrend.net a website that uses World Bank data sources, the current urban population of India is 471 million (2019) which is increasing at the rate of 2.33 per cent from the last year figure (See

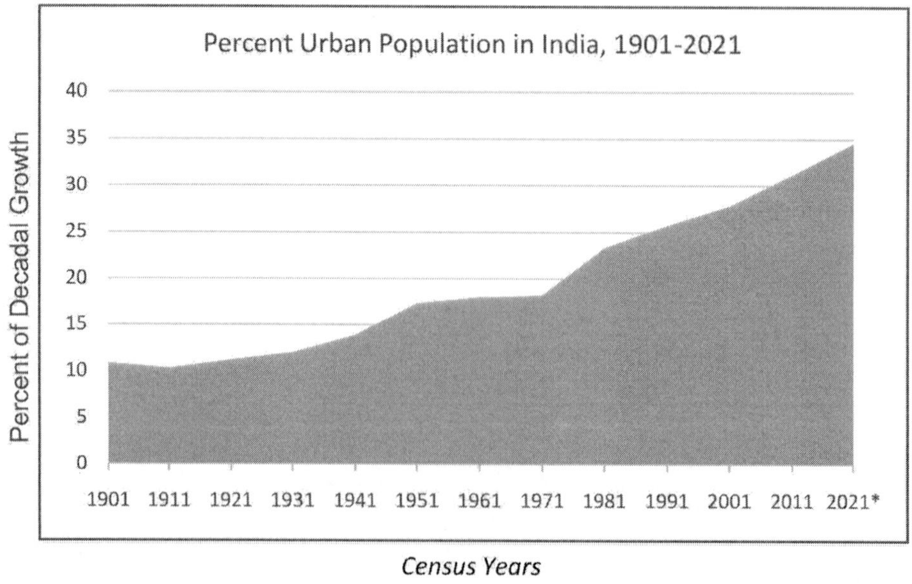

Figure 1:Per cent urban Population in India- 1901-2021

*Sources: Various census years- 1901 to 2011, *NCP projected population data for 2021*

Figure 2). These 471 million urban populations (34.47 per cent share) have already surpassed the total USA population. India's urban population is more than the total urban population of North America (304.76 million in 2019) and South America (368.40 million in 2019) respectively. However, the percentage urban population of North American and South America is more than 82.6 per cent and 85.5 per cent, respectively. Compared to Europe with 556.68 million urban population which is 74.5 per cent of the total population of Europe, India lags far behind in the levels of urbanization, while at the same time the total urban population of India is much bigger than both the American Continents respectively and is about 85 per cent of the total urban population of Europe. Indian's level of urbanization is comparable with the South Asian countries having 709.38 million urban population with 36.6 per cent urban pupation to the total population of this region, while India alone contributes 471 million urban population in this region. China's level of urbanization is 61 per cent with a total urban population of 875.07 million in 2019. Much of the urbanization will happen in South Asia and African continents as the countries in these regions are fast urbanizing. African urbanization is at present 44 per cent with a 587.74 million population. Compared to the fast-urbanizing countries of the world, India's progress is slow, however, given the pace of economic growth that emphasizes urban activities, it is likely that the increasing trends of urbanization will continue and India may reach 50 per cent levels of urbanization in 2050.

As far as the decadal rate of growth of the urban population is concerned, the 110 years data (see Table 1.) shows that it has reached a peak (46 per cent) during the 1981 census and has been declining till 2011 when it has reversed the declining decadal rate of growth with 31.8 per cent in 2011 compared to 31.74 per cent during 2001. It is during the last decade that the increase in the total urban population i.e. 91 million was greater than that of the total increase of the rural population of 90 million (Bhagat, 2011). This is called India's urban turn! National Commission on Population projects India's urban population as 35 per cent and 38 per cent for the years 2021 and 2031 respectively, it may reach up to 40 per cent by the year 2036. Thus, reaching a 50 per cent share of the urban population in India by 2050 may depend on various factors of urbanization in future.

The pace of urbanization and absolute number of urban population matter as far as the urban development in India is concerned. The large urban population in India has been considered both problem and a prospect at the same time. Besides, urbanization per se brings a lot of problems altogether. The followings are the major problems of urbanization in India.

Challenges of Urbanization in India

Rapid urban growth

The growth of the urban population is faster in India compared to many developed countries of the world, however, it is comparable to the South Asian countries. Rapid urban growth is taking place in the large cities while small and medium towns are not growing rather there has been a declining growth rate of the urban population, though there is no negative growth in these towns in the last century. The major issues of the rapid urban growth can be summarized in the following points below:

- The increase of 91.1 million persons to the urban population during 2001-2011 is not only the highest registered thus far but also higher than the increase of 90 million persons to the rural population.
- The annual exponential growth rate of 2.76% registered during 2001-2011 has reversed the declining trend observed during 1981-91 and 1991-2001.
- The rise in urban population growth rate has come during an era of a sharp decline in the natural growth rate showing that the push to urban population has come in from other sources i.e. rural to urban conversion and rural-urban migration.
- The number of metropolitan cities (1 million-plus) has risen sharply, from 35 to 53 during 2001-2011. They now account for 42.6 per cent of the total urban population. Likewise, class I cities (1 Lakh plus) now account for 70.2 per cent of the country's urban population. In 2001, this percentage was 68 per cent.
- This phenomenon is called metropolitisation in India as far as the political economy of urbanisation is concerned (Sharma and Abdul, 2006)

Urban Expansion, Urban Sprawl, and Agglomeration

Large metropolitan cities such as Delhi, Mumbai, Kolkata, Chennai, Bangalore, Hyderabad, Surat and Jaipur etc. are expanding spatially and promoting huge urban agglomerations around them. The built-up areas are increasing particularly in the last two decades. This may be the impact of globalization and liberalization of the Indian economy which promoted huge investment in these cities in real estate and other urban development activities. This has resulted in creating a huge demand for the labour market in these cities hence, they have shown rapid growth of population during the last two decades. However, these cities are not well prepared to tackle the explosion of urban population and many a time the urban services failed to provide adequate provisions of housing, water and sanitation, and decent urban living to all the citizens. The worst is the case in the periphery or peri-urban areas that results from this rapid urban expansion. The trunk infrastructures are missing in these areas and urban services are very poorly provided. Sometimes they are not governed directly by the urban local bodies and have problems with governance. Many of the peri-urban areas around the large cities are governed by the Panchayati Raj Institutions and they lack the basic urban amenities. The delay in granting the urban status to these peri-urban areas causes many issues in the short and long run. The metropolitan authorities of these areas lack the financial capacity and authority to deal with these emerging problems of urbanization. The following Figure 1. is the examples of the expansion of built-up areas in the last two decades of the selected cities- Mumbai and Delhi in India.

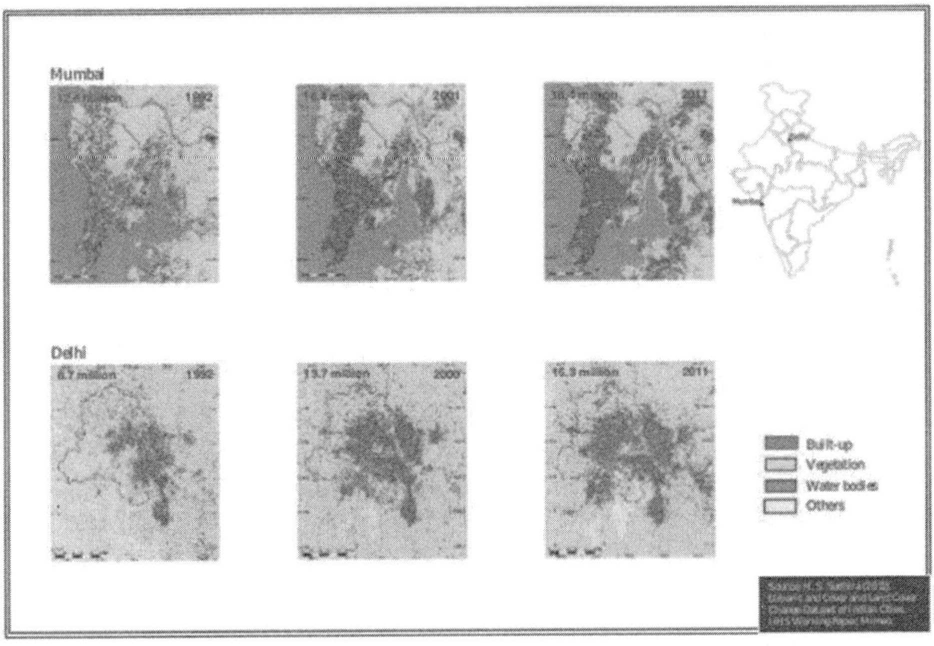

Figure 2. Change in the built-up area and land cover - Mumbai and Delhi 1992-2011

Informal urban Economy

The urban economy is highly informal. This means that the available labour force is accidentally employed in informal economic activities generally in large urban areas. The informal economy is defined (by International Labour Organisation- ILO) as the followings:

The informal sector is broadly characterised as consisting of units engaged in the production of goods or services with the primary objective of generating employment and incomes for the persons concerned. These units typically operate at a low level of organisation, with little or no division between labour and capital as factors of production and on a small scale. Labour relations - where they exist - are based mostly on casual employment, kinship or personal and social relations rather than contractual arrangements with formal guarantees.

Due to lack of contractual arrangement and formal guarantee of job security and other related social benefits, those workers employed in either a) self-employed in their enterprises or b) informal wage employment, suffer from employment-related risks and vulnerabilities. They do not receive legal and social protections

Box 1. Informal employment as a per cent of total, rural and urban employment

Countries by income level*	Total	Rural	Urban
World	61	80	44
Developing	90	90	79
Emerging	67	83	51
Developed	18	22	17

*2018 World Bank definitions based on country levels of gross income per capita.

Source: ILO, 2018

Box 2. Definitions recommended by NCEUS

Informal Sector: "The unorganized sector consists of all unincorporated private enterprises owned by individuals or households engaged in the sale and production of goods and services operated on a proprietary or partnership basis and with less than ten total workers".

Informal worker/employment: "Unorganized workers consist of those working in the unorganized sector or households, excluding regular workers with social security benefits provided by the employers and the workers in the formal sector without any employment and social security benefits provided by the employers".

Informal economy: The informal sector and its workers plus the informal workers in the formal sector constitute the informal economy. This is also called *shadow economy.*

(Sood, 2016). These workers generally are employed on weak employment conditions and low wage rates. ILO (2018) estimates that there is 44 per cent of workers informally employed in the urban areas of the world (see Box 1.), whereas developing countries employ 79 per cent of the workers. Urban areas in emerging countries provide 51 per cent informal employment while the developed nations offer only about 17 per cent of urban informal employment. It has been observed that the size of informal economic activities determined the inequality in the economy. The greater the informal economic activities in each country the more is the inequality, where there is a huge income gap between rich and poor.

The urban economy in developing countries particularly in India consists of informal sectors, informal employment, and informal activities (see Box 2). These are strongly related concepts and broadly linked with poverty, inequality and urban development (Mehta, 1985; Kundu, 1999). The total size of informal employment in India is 81 per cent as the latest ILO report in 2018. However, the Ministry of Labour and Employment, Government of India report based on the NSSO 68th round survey in 2011-12 suggest that there is 69.1 per cent of workers engaged in informal enterprises in non-farm activities. Urban informal employment is better than rural informal employment as the regularity of jobs is found more in urban areas. The major economic sectors that employ informal workers are non-manufacturing including construction industries, manufacturing - both household and non-household manufacturing and service sectors in the urban economy in India. The spatial manifestation of informal economy in urban areas are seen in terms of urban informal housing and slums.

Informal Settlement

They are characterized by dilapidated housing and poor living conditions and the lack of basic services such as water and sanitation. Formal definitions of slums in urban areas are given in Box 3.

The slums are defined by the Census of India and NSSO and also by the respective state government depending on the prevailing socio-economic characteristics of the region, its terrain and the conditions of housing.

- Slums are often related to the informal settlement in the cities with inadequate housing and related amenities such as water supply, sanitation, and other basic infrastructures. They are overcrowded and unhygienic. Slums are not a new phenomenon of urbanization. They

Box. 3 Definition of Slums by the Census 2011

i. All notified areas in a town or city notified as 'Slum' by State, UT Administration or Local Government, Housing and Slum Boards etc. under any Statute including a 'Slum Act' are considered as Notified Slum.

ii. All areas recognized as 'Slum' by State/Local Government, UT Administration, Housing and Slum Boards etc., which may have not been formally notified as slum under any statute are categorized as Recognized Slum.

iii. A compact area of at least 300 populations or about 60-70 households of poorly built congested tenements, in unhygienic environment usually with inadequate infrastructure and lacking in proper sanitary and drinking water facilities in the State/UT are categorized as Identified Slums.

Source: GOI, 2015

have been there during the phase of industrialization that led to urbanization during the colonial period. Though it provides only a solution for affordable urban living for the urban poor and those employed in the informal economy of the city, it may not be a sustainable solution to decent urban livings. The reason for slums in India is the rapid urban population explosion and the non-inclusive nature of urbanization (Kundu and Saraswati, 2012) that does not easily accommodate the distressed rural to urban migrants in the cities. On the other hand, poor quality of rural-urban migration leads to poor quality of urbanization (Bhagat, 1992). The status of slums in India can be summarised below based on a report of Government of India, 2015 (data used below are of Census, 2011):

- Out of 4041 statutory towns, 2613 cities/towns have reported slums in their jurisdictions;
- The slum population constitutes 22.4 per cent of the total population of the 2613 slums reporting towns;
- 19 census towns in NCT of Delhi and one census town in Uttar Pradesh have also reported slums;
- These towns are spread across 31 States/UTs of the country;
- Four States/UTs- Manipur, Daman & Diu, Dadra and Nagar Haveli and Lakshadweep have not reported any slum;
- The slum population constitutes 5.4 per cent of the total population of the country;
- While the urban slum population constitutes 17.4 per cent of the

total urban population.

Slums have been reported from all the major cities in India. The largest four metropolitan cities of India such as Mumbai, New Delhi, Kolkata and Chennai have reported slums in their municipal boundaries more than the national average figure of 17.4 per cent.

Table 2
The population of Slums in Four Largest Metropolitan Cities in India - 2011

Sl. No.	City	Total Population*	Slum Population	Per cent age slum Population
1	Chennai	4343645	819873	18.9
2	Delhi	9879172	1851231	18.7
3	Mumbai	11978450	6475440	54.1
4	Kolkata	4572876	1485309	32.5

*Population of City Municipal Corporation Area

Source: Census of India, 2011

Table 2. shows that Mumbai Municipal Corporation has reported the highest slum population of 54. 2 per cent of the total population of Mumbai City, followed by Kolkata, Chennai and Delhi. The associated problems of slums are manifolds. Apart from inadequate housing and poor living conditions,they suffer from various kinds of poverty including health and educational poverty. It has been reported that slums are major concentrations of crimes and other anti-social activities in Major cities in India (Prasad and Gupta, 2016; Srivastava and Singh, 1996). This is a failure of governance and neglect for the marginalized section of society. However, it has been observed that increasing urbanization has positive impacts on the reduction of urban poverty and slums. Large cities have been trying to reduce poverty and address the various problems of slums. Poverty reduction in urban areas has been faster than its rural counterpart. There are more serious issues that become threats to urban life particularly in large cities due to the rapid growth of urbanisation. They are urban environmental problems such as water and air pollutions.

Urban Environment Related Issues

The ambient air qualities of large cities are posing problems of health hazard to urban dwellers. Major cities of India report dangerous and poor ambient air quality daily. The case of Delhi's winter fog and air contamination in terms of Particulate Matter (PM 10, PM 5 and PM 2.5) and other air pollutants such as CO2; Sox2; and Nox2far accede the human tolerance level and resultant health problems such as lungs diseases and related complications are rampant during the season. The major causes of this poor-quality air in the large cities in India are emissions from vehicular mobility. The major share of CO_2 emissions in the cities comes from public and private vehicular pollutions. This has been also the major contributor to rising temperature in the cities. The ambient temperature of the cities across the seasons in India are always greater than rural hinterlands and this phenomenon is called urban heat island. The cities in a way are the urban heat islands causing massive health hazards to their citizens.

Another environmental problem of the city is water pollution. Cities in India have major sources of water either from rivers, lakes or underground water systems. The urban river basins show a greater amount of water pollutions in India. The water quality parameters such as pH, BoD, COD, Fecal Coliforms, Heavy Metals, hydrocarbons etc. are found in a deteriorating state in almost all the major rivers in the cities that are located near to it. Based on the Central Pollution Control (CPCB) datathe Comptroller and Auditor General (CAG), 2005 report says that the downstream of river water in urban areas is highly polluted, water samples from Yamuna River in Delhi show signs of contaminations that far accedes human use of Bathing Quality Standards (Agarwal 2016). The similar is the case of Sabarmati River in Ahmedabad where upstream water quality meets the Bathing Quality Standards, however, it is almost dead downstream (Bansal and Parthasarathy, 2020;). IDFC report quoting Central Pollution Control Board Data of 2008 shows that more than 900 cities (Class I and Class II) have only 32 per cent water treatment capacity. The remaining 68 per cent untreated urban water which is called "grey water" drains directly into their respective rivers and water bodies. Urban centres not only need to reduce the sources of both water and air pollutions, but also increase the water treatment capacity by installing Sewage Treatment Plants (STPs). This should be the priority of urban governance in India before it gets too late.

Urban poverty, equity and crimes

Urban poverty in India is defined as absolute poverty i.e. following the Head Count Ratio (HCR) of people below the poverty line living in urban areas. It considers the expenditure on food per capita per day to meet the daily requirement of calorific value of 2400 Kcal for the person residing in urban areas. At present,there is no data available for the estimated number of urban poor in India. The Ministry of Housing and Urban Poverty Alleviation, Government of India and UNDP released India-Urban Poverty Report in 2009, which uses the then Planning

Commission data on urban poverty estimates. According to this report, there is more than 80 million urban poor in India which accounts for 25 per cent of the total urban population in the year 2004-5 (using NSSO 61th Round Estimation).

As per the Times of India Report "Urban poverty in India is over 25 per cent; close to 81 million people in urban areas live below the poverty line. In India, the rural poor outnumber the urban poor. In the next decade urbanisation in India is set to reach 50 per cent."

Key messages of the report include:

- **Poverty in India has become urbanised**. Urban poverty in India is over 25 percent; some 81 million people live in urban areas on incomes that are below the poverty line. At the national level, rural poverty remains higher than urban poverty, but the gap is closing. By 2030, urbanisation in India is projected to reach 50 percent.
- **Migration towards urban centres has increased**, indicating that economic reforms have not been effective in creating jobs in small and medium towns as well as rural areas. Poverty was higher among rural to urban migrants, while the most successful migrants are those who move from one urban area to another.
- **Urban poverty poses different problems**. The nature of urban poverty poses distinct challenges for housing, water, sanitation, health, education, social security, livelihoods and the special needs of vulnerable groups such as women, children and the elderly.
- **Slum populations are increasing**. According to a 2001 census, India's cities have a slum population of 42.6 million (23.7 percent of the urban population). The majority (11.2 million) are in Maharashtra, whose capital city Mumbai is home to the Dharavi slum. While the slum population has increased, the number of slums has decreased - resulting in greater density.
- **Slum-dwellers lack access to basic services**. Most slum dwellers do not have access to clean water, sanitation and health care facilities. They face a constant threat of eviction, removal, confiscation of goods and have virtually no social security cover. Some 54 percent of urban slums do not have toilets; public facilities are unusable due to a lack of maintenance.

Lack of urban basis infrastructure- Water and Sanitation

There is a gross inadequacy of water and sanitation in the nonmetropolitan cities in India. The cases of water riots are also reported in the big cities and the perpetual problem of water supply in Chennai can be cited as an example. Less than 50 percent of the total population of India have access to safe drinking water (https://www.unicef.org/india/what-we-do/clean-drinking-water) and about half of the country's people still open defecate https://www.wateraid.org/us/media/indias-water-and-sanitation-crisis). This is the state of affairs of water and sanitation in India. The challenges are more in small and medium towns as far as water and sanitation are concerned. According to a report prepared by Urban Management Centre under the Performance Assessment System (PAS) Project, based at the Centre for Research and Development Foundation, CEPT University, Ahmedabad for the state of Gujarat, It is found that the coverage of toilets in all the municipalities are about 87 percent and in the slums, it is 82 percent. The coverage means that the per cent age of dwellings having independent toilet facilities in their premises and or have community toilets. While only about 60 percent of the total slums and non-slums are connected to the sewage connection. There is a long way to go to have total (100 per cent) water and sanitation coverage in all the towns and cities in India. Gujarat being 42 per cent urban provides only 130 LPCD of drinking water to its urban dwellers (PASS, 2016). These are the real challenges of rapid urbanisation in India that need the urgent attention of planners and policymakers before it moves to make its cities smart and intelligent.

Urban mobility- Congestion and Chaos

Urban mobility has huge challenges and opportunities in India. Nearly about half of the population either do not travel or go to work on foot. The Indian population in a nutshell are not mobile. According to the Census of India 2011, a considerable amount of people still prefer to walk or travel on a bicycle to go to their respective workplaces.

Public and private transports are being equally divided with 16 percent of the urban population travelling by their private vehicle and the same percentage of population availing public transport system. This is not a sustainable transport system in the country with around 400 million urban population. Some of the major issues of urban transport in India are - the rapid growth of private vehicles ownership largely (75%) two-wheeler, congestions in the city, deteriorating air quality; increasing greenhouse gas (GHG) emissions from the transport sector; increasing road accidents; and exploding growth in the number of private vehicles (largely motorcycles). Besides, there is a huge crisis of parking in the cities across the length and breadth of this country. These are the

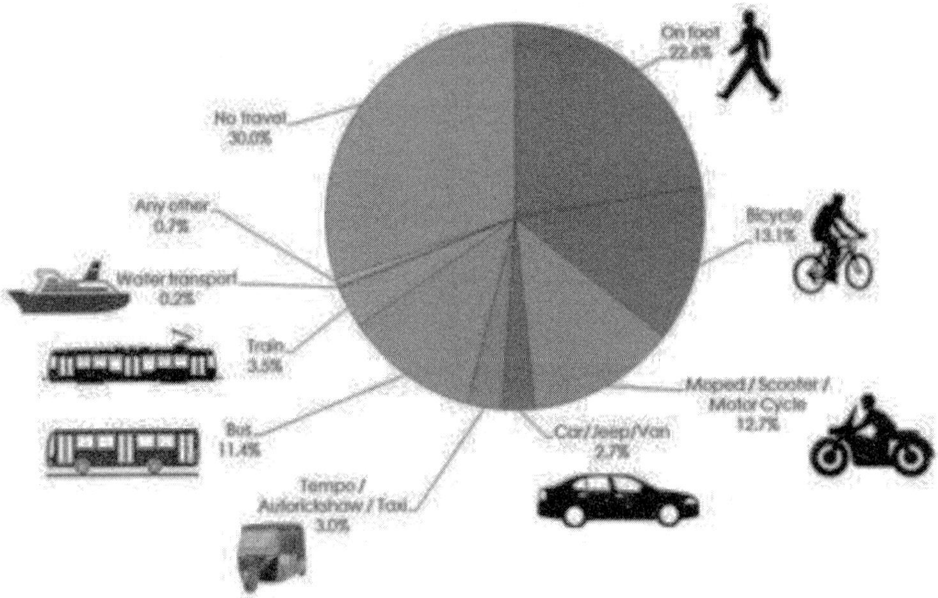

Source:https://www.intelligenttransport.com/transport-articles/21458/city-public-transportation-india/

detrimental factors of India's urbanisation and economic growth.

Conclusion

Urbanisation in India faces manifold challenges and issues related to urban expansion and rapid urban growth in big cities. The growth of small and medium towns are restricted and urban sprawl is happening in the big cities by either reorganization of the city's boundaries or by its natural growth couple with in-migrations from rural to urban largely in the big cities. Informal settlements are mushrooming in the cities that make the entire urban economy informal with less regular job opportunities, low levels of social security and lack of secured tenure making the large urban population vulnerable to various socio-economic conditions and climate-related disasters. The deteriorating urban environment coupled with lack of urban basic services such as water and sanitation gives rise to poor urban livings manifested with inequality, poverty and crimes in the urban centres in India. Urban mobility also faces challenges as to how to make the public transport system efficient in the cities, while 50 per cent of urban dwellers do not use any transport system, at the same time share of the public transport system is equal to those of the private transport system. Congestions and longer travel time in the cities make commuters' life difficult, however, parking in the cities has been the real challenge of urban mobility. The above issues and challenges are obvious at this stage of urban urbanization in India. The futuristic and sustainable urban planning need to consider these challenges and try to address the issues on a priority basis for making India's urban future more resilient, sustainable and livable.

References :

Aggrawal, M. (2016). High air and water pollution in India's 41 tier-II cities, Mint, Updated: 28 Apr 2016, https://www.livemint.com/Politics/r8blEXdbvNDhWmoylH7b5K/High-air-and-water-pollution-in-Indias-41-tierII-cities.htmlRetrieved on 31 March 2021

Aijaz, R. (2021). Managing India's urban transition in 2021, Indian and the world in 2021, urban future, Observer Research Foundation, New Delhi, accessed online on 26.03.2021

https://www.orfonline.org/expert-speak/managing-india-urban-transition-2021/

Bhagat, R.B., (2011). Margining pattern of urbanization in India, Economic & Political Weekly August 20, 2011 vol XLVI No. 34

GOI, (2014) Report on Employment in Informal Sector and Conditions of Informal Employment Volume IV Government of India, Labour Bureau, Chandigarh

GOI, (2015) Slums in India: A Statistical Compendium, Ministry of Housing and Urban Affairs, National Building Organization, New Delhi

https://www.worldometers.info/world-population/europe-population/ Retrieved on 26.03.2021

https://www.worldometers.info/world-population/south-america-population/ Retrieved on 26.03.2021

https://www.worldometers.info/world-population/northern-america-population/ Retrieved on 26.03.2021

https://www.macrotrends.net/countries/IND/india/urban-population/ Retrieved on 26.03.2021

https://www.worldometers.info/world-population/southern-asia-population/ Retrieved on 26.03.2021

https://www.worldometers.info/world-population/china-population/ Retrieved on 26.03.2021 https://www.worldometers.info/world-population/africa-population/ Retrieved on 26.03.2021

https://www.worldometers.info/world-population/oceania-population/ Retrieved on 26.03.2021

https://www.worldometers.info/world-population/japan-population/ Retrieved on 26.03.2021

ILO, (2018). Women and Men in the Informal Economy: A Statistical Picture (Third edition) / International Labour Office - Geneva: https://www.wiego.org/publications/women-and-men-informal-economy-statistical-picture-3rd-edition, Retrieved on 28-03-2021

Kundu, A and Saraswati, L.R., (2012) Migration and Exclusionary Urbanisation in India, Economic & Political Weekly, Vol. 47, Nos. 26 & 27

Kundu, A. (2014). Exclusionary Growth, Poverty and India's Emerging Urban Structure. Social Change, 44(4), 541-566. doi:10.1177/0049085714548538

Kundu, Amitabh (1999). Urban Informal Sector in India: Macro Trends and Policy Perspectives, Issues in Development, Discussion Paper No. 25. ILO, Development Policies Department, Geneva

Mehta, M. (1985). Urban Informal Sector: Concepts, Indian Evidence and Policy Implications. Economic and Political Weekly, Vol. 20 No.8, pp326-332. Retrieved March 28, 2021, from http://www.jstor.org/stable/437411

Prasad, R and Gupta, N., (2016). Problems and Prospects of Slums in India, International Journal of Multidisciplinary Approach and Studies, Volume 03, No.3, May-June 2016

Sharma, A., (2019) Cities of the Poor: A view on Urban Poverty in India, November 21, 2019, The Times of India, https://timesofindia.indiatimes.com/blogs/in-the-name-of-development/cities-of-the-poor-a-view-on-urban-poverty-in-india/ Retrieved on 31 March 2021

Sharma, R.N., and Abdul, S., (2006). Metropolitisation of Indian Economy: Lessons in Urban Development in Ed. Sisodiya, A.S. And Rao, N.J., Indian Reality the Changing Landscape, ICFAI University Press

Sood, A. (2016). Urbanisation and the Informal Economy. Sociology of Urban Transformations. E-PG Pathshala: http://epgp.inflibnet.ac.in/

index.php Retrieved on 28-03-2021

Srivastava, A., & Singh, R. C. (1996). Slums and associated problems: a case study of Bhilai, an industrial city, India. International Journal of Environmental Studies, 50(1), 51-60. doi:10.1080/00207239608711038

www.macrotrends.net/countries/IND/india/urban-population'>India Urban Population 1960-2021. Retrieved 2021-03-26.

3

Urban-Rural Linkages and Environmental Issues

Abha Lakshmi Singh
abhalakshmisingh@yahoo.com

This study has drawn the attention towards the often-neglected subject- 'urban rural linkages and environmental issues. The urban development and urban environment cannot be seen in isolation from its surrounding rural areas. This points towards the fact that urban-rural relations do exists and day by day with galloping urbanization, the intensity of these linkages have increased and so have the environmental consequences. But the environmental issues which result from these linkages are often overlooked and ignored. This study comprehends the nature and intensity of urban-rural relations and the resultant environmental issues as a whole in a comprehensive manner.

The work seeks to:

- Improve our understanding of how urban-rural linkages/interactions affect the environment,
- How the linkages are not all positive or beneficial to both ends of the spectrum,
- How the environmental issues imposed on the surrounding rural areas are due to urban activities which involve either,
    - o Use of rural land
    - o Urban demands of rural resources or
    - o Disposal of waste on rural land or water

- The environmental issues should not be ignored but be given proper consideration by policy makers.

The study focuses on the fact that environmental issues and changes taking place in the surrounding rural areas are linked to urban-rural interactions that involve flows of people, goods, income, natural resources and wastes. The analysis of flows shaping and reshaping the use of environmental resources and ecological services show how the three are interconnected yet varied levels drive them,

- At the first level change is driven by local conditions including for instance the competition between urban development and agriculture for land or increasing pressure of extractive activities as a response to the city demand for building materials.
- At the second level change is driven by regional and national conditions such as the promotion of industrialization.
- At the third level change is driven by international conditions such as falling prices of export crops that increases the migration of impoverished farmers to the peri-urban interface in search of alternative livelihood strategies.

Environmental issues often occur without people realising as it is happening. When change occurs.

- The environment is damaged
- The poorer people suffer the most

This is because,

- They rely heavily upon the depleting resources such as water, firewood and local resources.
- Their health and living standards often suffer because they live close to the sources of pollution.

Most environmental issues/changes occur in three ways,

- **Changes in landuse** for example from agriculture to residential, commercial, industrial etc.
- **Changes in use of natural resources** for example deforestation, soil erosion, water depletion etc.
- **Changes in the generation of waste** for example increased solid and liquid waste resulting in solid and water pollution

The environmental issues have been ignored by policy makers.

Increasing production and consumption is environmentally ambiguous. It tends to increase both resource demands and waste generation. Superficially, the combination of economic growth and urbanization might head to lead inevitably to greater and more concentrated environmental impacts. The six case studies presented demonstrates that day by day the linkages/interactions are becoming even more intensive and an important component.

It is now widely recognized that there exists an economic, social and environmental interdependence between urban and rural areas and there is need for a balanced and mutually supportive approach to the development of the two areas. The discrete consideration of rural development as completely distinct from urban development is no longer valid and is replaced by the urban-rural linkages development approach. Indications are that the urban-rural interactions are set to continue and escalate in the coming decades. The linkages are not only in topological sense of neighbourhood. **A multidimensional interdisciplinary approach is required for improving the development. Only a sustainable use of natural resources with thematic linkages of economic, social and environmental urban-rural inter relations can provide the basic for human-wellbeing. To achieve sustainable development the Ecosystem Approach 'a strategy for integrated management of land and water and natural resources that promotes conservation and sustainable use in an equitable way' should be used on cities and its hinterland.**

The study points towards the fact that the issue of urban rural linkages and resultant environmental consequences include a number of key issues such as,

- Environmental impacts of physical expansion of urban areas on the surrounding rural areas.
- Environmental impacts of urban demands for resources in the surrounding rural areas.
- Environmental impacts generated by the city-based activities in the surrounding rural areas.

Increasing number of people (in-migrants from the surrounding villages and small urban centres; out migrants from the city) are moving into the areas around Aligarh city; consequently, the demand for land and accommodation in these areas is steadily rising and so is the city expanding in the surrounding rural areas. As the city population has increased from 480,520 in 1991 to 912,388 in 2011 and by 2021 it is expected to be 1,249,352. So has the city area grown from 36.70 sq.km in 1991 to 78.23 sq.km in 2011 and by 2021 it is expected to be 114.70sq.km.

The physical expansion of urban areas is always over rural areas and so the most obvious environmental change is that associated with the expansion of the built up area and the immediate environmental impact of urban activities that develop there. The expanding urban areas transforms not only the land that becomes urbanized but also much larger areas as has been seen from the changes in rural landscape and ecology determined by the productive activities that respond to urban-based demand for inputs e.g. raw materials, goods and services. Large demands are made on the surrounding areas of cities for building materials and landfill, as a result of the construction of buildings, roads, industries and other components of urban fabric. Many of the urban generated solid and liquid wastes impacts surrounding areas-often disposed on open-air sites with no provision for protecting surrounding land and water from contamination.

If well managed, the interactions between urban and rural areas will be the basis for a balanced regional development and also will be economically, socially and environmentally sustainable. Local development is increasingly associated with decentralisation processes, on the assumption that local government is closure to citizens-meaning that both are accountable to them and that it has a better understanding of local needs and priorities. With regard to urban-rural linkages, local government can play an important role in facilitating positive interactions and limiting negative exchange.

- The management of natural resources and wastes in an important area of local government intervention.
- National level policies have an important role with respect to land and land ownership in both urban and rural areas.

In short, **understanding urban-rural linkages matter with serious concern as they provide the basis for measures that can improve both urban and rural livelihood and environment**. Ignoring them means that important opportunities will be lost and, in many cases, it will also contribute to poor and marginal people's hardship. There are urban initiatives that can reduce ecological damage to rural areas and support regional development. However, with an edge to urban-centric approach in the present scenario, such initiatives are unlikely to be given the priority they deserve.

4

Geomorphic Characterisation of a Part of the Ganga River System between Varanasi and Munger

Atul Aditya Pandey and Ramesh Shukla
atulgpandey@gmail.com

Abstract

River erosion and rejuvenation must take into account the holistic approach in investigations. It requires in-depth evaluation of the river morphology, geology, tectonics, climate flow-ecology, aquatic biota, landuse, livelihoods, agriculture wetlands, forests industries etc. in the basin. The objective of this paper is to define the geomorphic condition of a part of the Ganga river system between Varanasi and Munger. The project was carried out as a part of the Ganga River Basin Management Plan with the aim "maintenance and restoration of wholesomeness of Ganga system and improvement of its ecological health with due regard to the conflict of interest in water uses in entire river basin. An in-depth evaluation of the terrain was performed using multispectral Landsat images. SRTM DEM data have been used to delineate valley margin. Broadly two geomorphic units have been mapped - in-stream and flood plain. The most common in-stream geomorphic units are sandbars, mid-channel bars, point bars and lateral bars. The two major elements identified in the flood plain are meander scroll and flood channel. The geomorphic maps have been used to calculate different morphometric parameters such as sinuosity; braid channel ratio, bar area etc. Based on the characteristic geomorphic features both in-stream and flood plain, the stretch is divided into four Styles-Varanasi to Dheena, Dheena to Chhapra, Chhapra to Mokama and Mokama to Munger.

Introduction:

A river is a complex multidimensional resource. It requires in-depth evaluation of the terrain with respect to river morphology, geology, tectonics, climate, environment, landuse, wetland, forests, agriculture, aquatic biota, ecology etc in the basin.

Ganga River Basin Environmental Management Plan sponsored by the Ministry of Environment and Forests and climate change, Government of India aimed at the "maintenance and restoration of wholesomeness of the Ganga system and improvement of its ecological health with due regard to conflict of interests in water uses in entire river basin. The fluvial geomorphology component of the project defined the geomorphic condition of the Ganga river system in different reaches which helped to understand the hydrology -geomorphology -ecology linkage for developing a sustainable river management programme. The objectives of the geomorphological investigations is to prepare and compile geomorphic map of the Ganga river and classification of the reaches in terms of their geomorphic condition.

Methodology:

Multispectral Landsat images have been used for mapping geomorphic features and SRTM DEM data to delineate valley margin. The geomorphic units that have been mapped for the river Ganga are Valley margin, active flood plain and channel belt each characterized by several geomorphic features. These geomorphic maps for the river have been used to calculate different parameters such as sinuosity, braid channel ratio, the bar area etc and has been used for characterizing the River Styles. Bed material character, types of riparian vegetation were recorded during field visits along the river.

Results and Discussion

In order to characterize the stretches of Ganga River and divide it into geomorphologically significant and distinctive portions, the guidelines outlined in the River Style frame work (Brierly and Fryirs, 2005) have been used. This has formed the basis for differentiating the Ganga River into distinct River Styles. The River Style framework has a nested hierarchical top down approach and is arranged in three scales:

1. Landscape/valley setting
2. Channel Confinement
3. Flood Plain Boundation

Based on the characteristic geomorphic features both in-stream and flood plain the stretch between Varanasi and Munger is divided into four styles. The river style one from Varanasi to Dheena, style two from Dheena to Chhapra, style three from chhapra to Mokama and the style four from Mokama to Munger.

The detail of each of the style is described in the tables given below:

RIVER STYLE-1: Varanasi to Dheena

Defining Attributes: The river style stretches for 92 km, trending NE - SW to E-W and is set within an alluvial setting. The channel which is single throughout the style, oscillates from left to right bank of valley margin. The floodplain is falling alternatively on the reverse flanks of the channel and being widest in the centre. Maximum valley width 20 km in the central part. There are number of lateral bars and few mid channel and point bars. Sinuosity of the channel 1.67 reflecting its meandering nature.

DETAILS OF ANALYSIS

Data used: Geomorphic and active floodplain mapped from Landsat 4-5TM imagery; Valley margin maps from SRTM.

RIVER CHARACTER

Landscape Setting	The channel flowing within an alluvial valley setting north of the Son river joining just at the eastern end of this river style.
Valley-Setting	Alluvial valley setting extending on either side of the river; valley width - 7.4 km- 20 km at the eastern end.
Valley Morphology	Average valley width 20 km, (max. 21.57 and min. 6.94), Valley showing a maximum width of 21.57 near dheena and it shows a minimum width of 6.94 near Varanasi.
Channel confinement	The channel is mostly unconfined. At places Kuchman it is confined on the southern bank. Further east the channel is semi confined along with the meander (M4) near Dheena (Reach 3 of the river style)

Active Floodplain Characteristics	Average floodplain width is 5.67 km (Max 12.96 km and min is 5.67 km). The floodplain is falling alternatively on the reverse flanks of the channel and being widest in the centre.
Channel Characteristics	Sinuosity (P): 1.67 Braid channel ratio (B): Total bar area = 40.09 sq.Km. Side bar: 22.66 sq.Km; length:1.95 km- 7.08 km; width: 0.17km-1.27 km. Alluvial Island: nil. Mid channel bars: 7.00 sq.Km, Length: 2.15 km-5.95 km, Width: 0.35 km-0.96 km. Point bars: 10.44 sq.Km, Length:1.60 km-5.60 km; width: 0.47km- 1.73 km. Confluence bars: nil. Flood channel: 5 in number, varying length from 1.23 km to 4.23 km. Meandering parameters: 4 meanders Radius of curvature (rc): 1.40 km-2.80 km. Wavelength (L): 5.71 km-18.00 km. Amplitude: 3.10 km-6.80 km.
Diagnostic geomorphic features (geometry sedimentology)	Channel Belt - Alluvial A couple of mid channel bars and 11 side bars; 4 meanders. Flood plain The floodplain is falling alternatively on the reverse flanks of the channel and being widest in the centre. Few flood channels contribute to water and sediments during monsoon.

Land Use Land cover	Large part of flood plain is used as cultivated land.
association	Channel belt

RIVER BEHAVIOUR

Low flow stage: River channel mostly uniforms in width, midchannel bars are exposed at four to five places.

Bank full stage: The rivers attain bankfull stage during monsoon, the sidebars and point bars are submerged.

Overbank stage: The river flood water submerges left and right bank alternatively.

RIVER STYLE-2 :Dheena to Chapra

Defining Attributes: The river style stretches for 195.02 kms and is set within an alluvial setting, trending east - west, maximum valley width - 35.80 kms near Ballia, Active flood plain on both sides, at one or two places, the channel flows along the valley margin, Meander scrolls and bow lakes are prominent along the southern side of the river. Ghagra a major tributaries trending NW - SE joins the main channel west of Patna. The floodplain is wide all along. There are number of side bars and few mid channel and point bars.

DETAILS OF ANALYSIS

Data used: Geomorphic and active floodplain mapped from Landsat 4-5TM imagery; Valley margin maps from SRTM.

RIVER CHARACTER

Landscape Setting	Valley setting is between the Son valley in south east and Ghaghra in the NE.
Valley-Setting	Alluvial valley setting extending on either side of the river; valley width - 9.6 km- 35.80 km at the eastern end.
Valley Morphology	Average valley width is 25.06 km, (max 35.80 km min 11.97km),
	Valley showing maximum width of 35.80 km near Ballia and minimum width near Chapra.
Channel confinement	Unconfined channel in the entire river style from Dheena to Chhapra.

Active Floodplain Characteristics	Average floodplain width is 17.20. (max 25.48 km and min is 7.45 km) Active flood plain on both sides. The floodplain is wide all along.
Channel Characteristics	Sinuosity (P): 1.09-1.81. Braid channel ratio (B): Total bar area = 158.78 sq.Km. Side bar: 59.73 sq.Km; length:1.52 km- 12.40 km; width: 0.25km-0.92 km. Alluvial Island: 64.13 sq.Km, length: 5.63 km-11.02 km; width: 2.90 km-5.10 km. Mid channel bars: 18.12sq.Km, Length: 1.34 km- 5.56 km, Width: 0.42 km - 1.15 km. Point bars: 11.89sq.Km, Length:2.14 km- 5.10 km; width: 0.61 km- 1.59 km. Confluence bars: 4.86sq.Km, length:1.66 km-3.59 km; width: 0.42 km- 1.77 km. Flood channel: 1 in number with length of 2.82 km. 1 chute channel with length of 3.48 km. Ghagara is the major tributary in this stretch. Meandering parameters: 5 meanders Radius of curvature (rc): 2.07 km-5.10 km. Wavelength (L): 7.20 km- 14.26 km. Amplitude: 2.85 km- 7.10 km.
Diagnostic geomorphic features (geometry sedimentology)	Channel Belt - Alluvial 3 alluvial island; one in the middle and the other two at the eastern end of this river style. Mid channel bars are distributed uniformly all along the river length.

	few meanders.
	Flood plain
	Active flood plain present on both side of the almost parallel to the stretch of the river.
	Major portion of the flood plain is characterised by meander belt having prominent meander scrolls.
	Several flood channels contribute water and sediments during monsoon.
Land Use Land cover association	Floodplain Large part of flood plain is used as cultivated land.
	I belt
	Alluvial islands are used for agricultural activities during low flows

RIVER BEHAVIOUR

Low flow stage: River channel mostly uniforms in width, midchannel bars are exposed at eight to nine places.

Bank full stage: The rivers attain bank full I stage during monsoon, the sidebars and point bars are submerged.

Overbank stage: The river flood water submerges left and right bank alternatively.

RIVER STYLE-3: Chapra to Mokama

Defining Attributes: The river style stretches for 146.90 km and is trending E-W and is set within an alluvial valley setting. It attains a maximum valley width of 20.0 kms across Patna and minimum of 8.10km d/s Gandak confluence. Active flood plain is mainly limited to left bank while the right bank is devoid of flood plain because of a southern embankment in the stretch of Patna Urban area upto the confluence of river Punpun form south. The channel has sinuosity of 1.21. It is characterized by presence of mid channel bars (31) and large alluvial islands (max, length - 34km, width - 12.5km). Alluvial islands seem to have gained prominence in width/ area d/s of confluence of Ghaghra and Gandak from north and son from south.

DETAILS OF ANALYSIS

Data used: Geomorphic and active floodplain mapped from Landsat 4-5TM imagery; Valley margin maps from SRTM.

RIVER CHARACTER

Landscape Setting	The river style stretches for 146.90 km and is trending E-W and is set within an alluvial valley setting. The floodplain is wide in the centre. There are numbers of lateral bars and few mid channel and point bars.
Valley-Setting	Alluvial valley setting extending on either side of the river; valley width - 20.0 km- 8.10 km.
Valley Morphology	Average valley width is 14.43 km, (max. 20.0 km min 8.10 km), Valley showing maximum width of 20 km near Patna and minimum of 8.10 km near Gandak confluence.
Channel confinement	Partly confined in the entire river style from Chhapra to Mokama.
Active Floodplain characteristics	Average floodplain width is 12.57 km. (max. 18.64 km and min width is 5.22 km). Active flood plan is mainly limited to left bank while the right bank is devoid plain because of a southern embankment in the stretch of Patna Urban area upto the confluence of river Punpun from south.
Channel Characteristics	Sinuosity (P): 1.21. Braid channel ratio (B): Total bar area = 677.94 sq.Km. Side bar: 79.46 sq.Km, length: 2.20 km-12.80 km; width: 0.66km-3.29 km. Alluvial Island: 535.29 sq. Km, length: 5.02 km-35.40 km; width: 2.47km- 11.72 km. Mid channel bars: 44.31sq.Km, length: 0.36 km-

4.92 km; width: 0.20 km-2.53 km.

Point bars: 13.39sq. Km, length: 2.50 km-5.17 km; width: 0.55 km-01.60 km

Confluence bars: 5.49sq.Km, length: 2.69 km-4.12 km; width: 1.30 km-1.32 km.

Flood channels: 5 in number, varying length from 1.93 km to 14.07 km.

4 chute channels of varying length from 6.37 km to 14.11 km.

Meandering parameters:

3 meanders

Radius of curvature (rc): 1.92 km-3.32 km.

Wavelength (L): 5.25 km-8.65 km.

Amplitude: 2.52 km-4.88 km.

Diagnostic geomorphic features (geometry, sedimentology)

Channel belt- alluvial
Alluvial Island cover more than 75% of the channel belt.

Couple off mid channel bars are distributed at the eastern end of the river style.

Few meanders.

Flood plain

Active flood plain present on northern side of the river almost parallel to its length.

Several flood channels and few chute channels.

Land use Land cover association

Flood plain
Large part of flood plain is used as cultivated land.

Channel belt

Alluvial Islands are used for agriculture throughout the year; side bars are also used for agriculture during low flows.

RIVER BEHAVIOUR

RIVER STYLE-4 : Mokama to Munger

Defining Attributes: The river style 71.49 km long and trending E-W is set within alluvial valley setting. It attains width 25.50km (max.) at further east of Begusarai. Active floods plain is present on both sides of the river with varying width, the channel characterized by sinuosity 1.35.

DETAILS OF ANALYSIS

Data used: Geomorphic and active floodplain mapped from Landsat 4-5TM imagery; Valley margin maps from SRTM.

RIVER CHARACTER

Landscape Setting	Alluvial valley filled
Valley-Setting	Alluvial valley setting extending on either side of the river; valley width - 11.50 km- 25.50 km at the eastern end.
Valley Morphology	Average valley width is 14.43 km, (max. 25.50 km min 11.50 km), Valley showing maximum width of 25.50 km east of Begusarai and shows minimum width near Munger.
Channel confinement	Mostly Unconfined channel; partially confined near Munger.
Active Floodplain characteristics	Average floodplain width is 20.44 km. (max. 20.43 km and min is 6.83 km). Active flood plan is present on both sides of the river with varying width.
Channel Characteristics	Sinuosity (P): 1.35. Braid channel ratio (B): Total bar area = 122.55 sq.Km. Side bar: 15.49sq.Km, length: 5.02 km-9.58 km; width: 1.30 km-2.39 km. Alluvial Island: 13.05 sq. Km, length: 10.00 km; width: 1.75 km. Mid channel bars: 6.00sq.Km, length: 1.10 km-

	4.60 km; width: 0.22 km-1.72 km. Point bars: 87.12sq. Km, length: 1.35 km-9.58 km; width: 0.35 km-4.20 km Meandering parameters: 3 meanders Radius of curvature (rc): 2.52 km-3.45 km. Wavelength (L): 5.55 km-10.50 km. Amplitude: 2.85 km-4.95 km.
Diagnostic geomorphic features (geometry, sedimentology)	Channel belt- alluvial Point bars distributed all along the river style. Few mid channel bar and one alluvial island 3 meanders Flood plain Active flood plain present on both side of the river lying almost parallel. Flood channel are prominent.
Land use Land cover association	Flood plain Large part of flood plain is used as cultivated land. Channel belt Alluvial Islands, side bars and few point bars are used for agriculture.

RIVER BEHAVIOUR

Low flow stage: River channel width decreases, side bars are exposed, flood channel dries up.

Bank full stage: Side bar and point are submerged. The flood channel supply water/ sediments to the active flood plain.

Overbank stage: River floods both the banks regularly, the flood plain is inundated providing lateral connectivity along both the banks.

Conclusions:

The Ganga River displays morphological diversity from Varanasi to

Munger. There are four styles each exhibiting its own characters. This stretch of the river lies in the alluvial setting and include confluence of major tributaries like Son, Ghaghra and Gandak. The river style one in the valley is characterised by abundant meander scars and sinuous channel suggesting a dynamic regime where meander migration is a common mechanism of river dynamics. Downstream Chapra the valley interfluves setting takes over and the most diagnostic feature in this style is the presence of large alluvial islands and branching channels. Over flooding along both banks of the river is common and an aggradational regime dominate this style. The lower reaches of Ganga river have a distinct river style as the southern bank of the river coincides with the craton margin. The diagnostic features of this style include wide valley and flood plains, large patches of abandoned meander belt and alluvial islands. The river style frame work highlights the geomorphic diversity along Ganga.

References:

Brierley, Gary and Fryirs, Kirstie (2005) Geomorphology and River Management Applications of the River Styles framework, Blackwell Publishing, Oxford, UK

Report submitted to the Ministry of Environment and Forests by Prof. Rajiv Sinha, Team Leader Fluvial Geomorphology Group, Ganga River Basin Environmental Management Plan, 2005.

5

Aspects of Urban Geological Studies w.r.t Patna Urban Agglomeration

Akhouri Bishwapriya, Priya Upadhyay, Abhishek Kumar Chaurasia, Shikha Dagar, G.S. Gokul and Sandeep Kumar

An area of about 700 km^2 covering parts of toposheet Nos. 72 C/14, 72 G/2, 3 & 6 on 1:50,000 scale, in Patna District, Bihar was taken up for Geo-environmental studies during 2018-19 & 2019-20.

The study of geo-environmental aspects viz. Geology, Geomorphology, Land Use and Land Cover, Pedology and Groundwater etc. are key indicators to urban development and their understanding as a direct bearing to the wellbeing of its subjects is important. The urban development plan of Patna district has gained importance in the wake of smart city development project of Governmentof India. Being

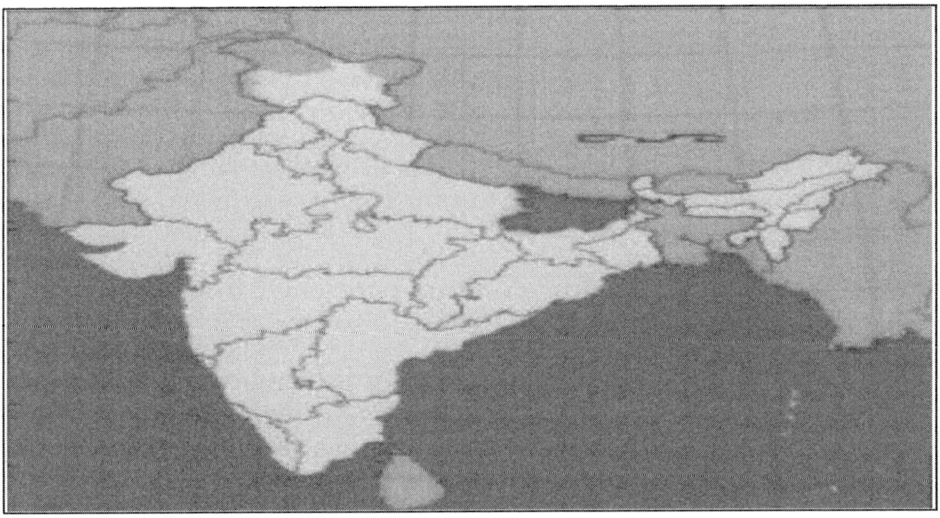

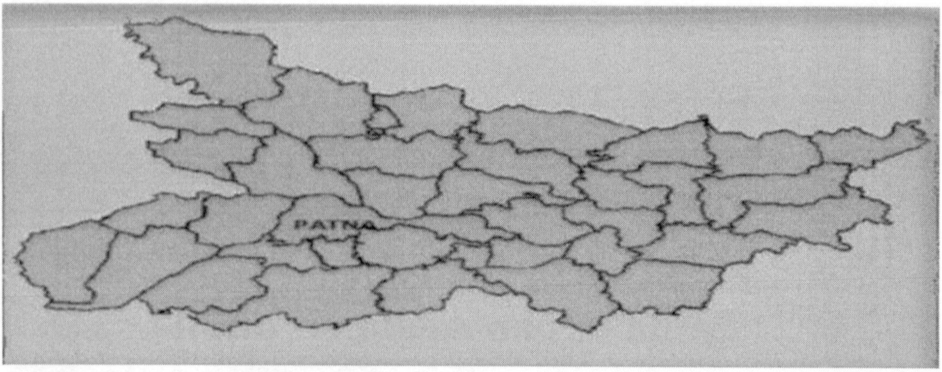

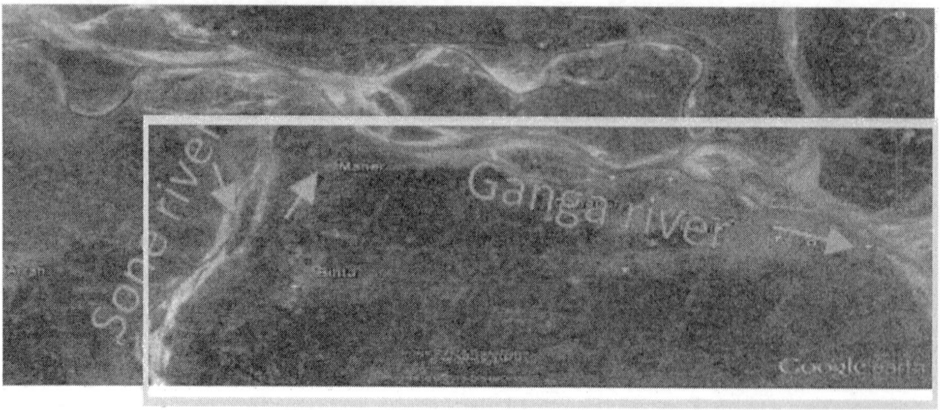

the second most populated state, there has always been a demographic pressure onthe state capital for better infrastructure facilities. But with growing needs, the commensurate status of naturally occurring resources, environmental hazards (seismic vulnerability, floods, contaminants in ground water) along with identification of factors of sustainable development isimportant. The expansion is thus inevitable and therefore the bearing of environmental concerns needs to be addressed

The study area is represented by Quaternary sediments ranging in age from Holocene to Recent. The sediment pile can be divided into two formations

Fatwa Formation (Ramgarh/Vaishali), represented by variegateds and, silt and clay

Diara Formation (Durgauti), represented by gray, unoxidised micaceous sand and silt with minor gravels, sometimes exhibiting alternating sequence of micaceous sand and silt with occasional clay and also includes the active flood plain deposits.

Geological Map of Parts of Patna, Bihar, India (T. No. 72C/14, 72G/2, 72G/3 & 72G/6)

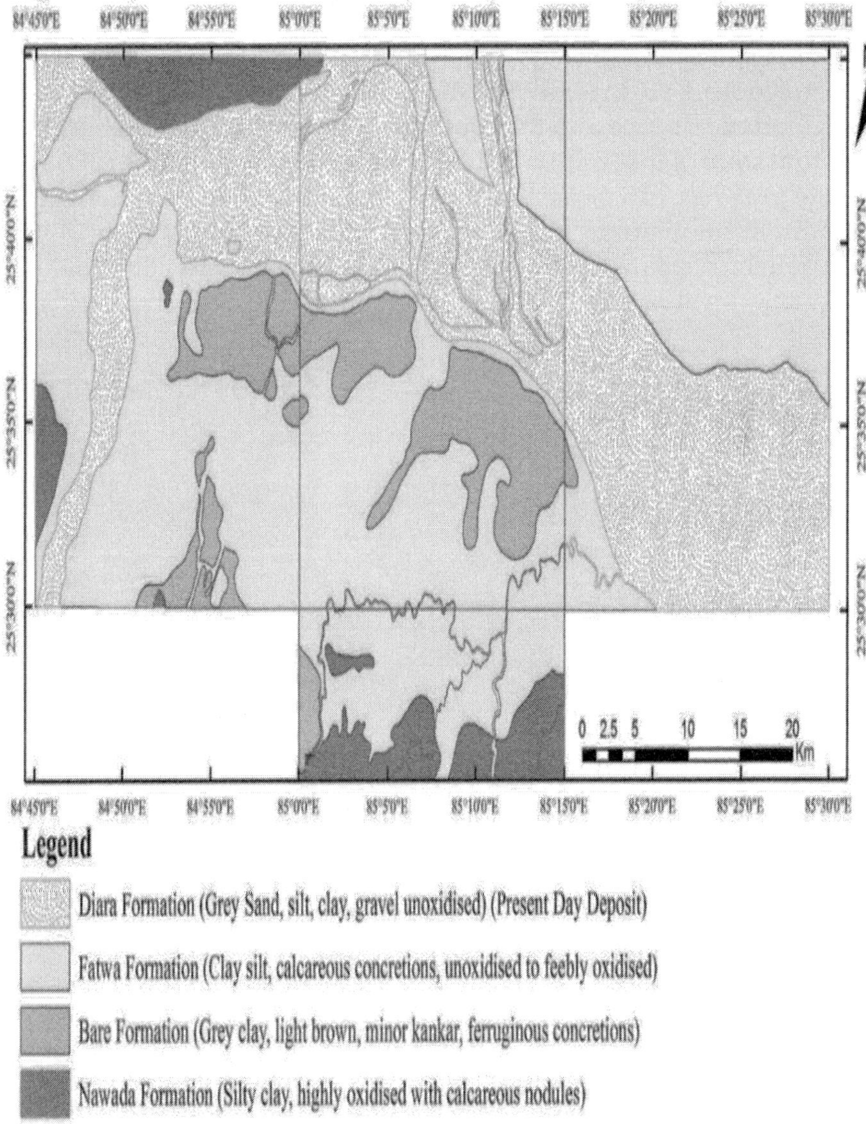

This work includes new set of inputs as envisaged in the scope of work, compilation of earlier workcarried out by Geological Survey of India, integrating requisite data and information from peer organisations like Central Ground water Board, Pollution Control Board, Central Water Commission, India Meterological Department etc. to arrive at requisite goals and objectives as envisaged.

In the scope of work, emphasis has been given to the following.

a) The impact of natural drainage: Patna is flanked to its north by Ganga river, to its south by Punpun river and to its west by Son river. These are principal higher order drainages with a potential to swell and inundate the adjoining low lying areas during peak discharge. They are also subjected to lateral oscillation owing to sediment influx, accommodation space and fluctuating discharge regime indicating their perpetual state of instability. Studies carried out in segments of Sonriver indicate presense of a persistent brown colour eroded basement of Older Alluvium at an average depth of 5 to 8 m b.g.l. This is invariably and ubiquitously overlain by silty to silty loam inlevee and silt plain, clayey to silty clay in flood basinal area and sandy litho-package in palaeo-

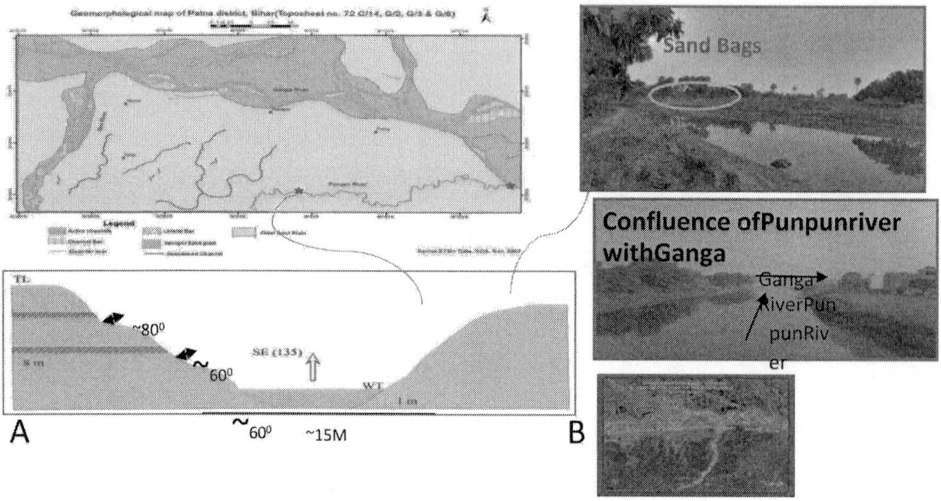

channels of Son River. Thus two generation of flood plains viz. Older Flood Plain and Present Flood Plain has resulted from the combined inter-fluvial activityof Ganga-Son-Punpun river system.

Punpun river is also guided by an abrupt change in flow direction (west to east from south to north) around the southern periphery of toposheet 72G/2 and is a threat to adjoining areas in terms off looding. The river through transverse sections in vulnerable stretches has been studied and findings included in the report.

The Ganga river is delimited on its southern bank by a raised levee but that diminishes between Danapur and Maner and likewise between periphery of toposheet 72G/2 upto its confluence with Punpun river. Typical sections have been studied and the status duly documented.

b) Pedological study: The pedological study has yielded data on physical and chemical components. Samples on grid pattern representing the

Land Use Land Cover Map of parts of Patna, Bihar, India (T. N. 72C/14, 72G/2, G/3 & G/6)

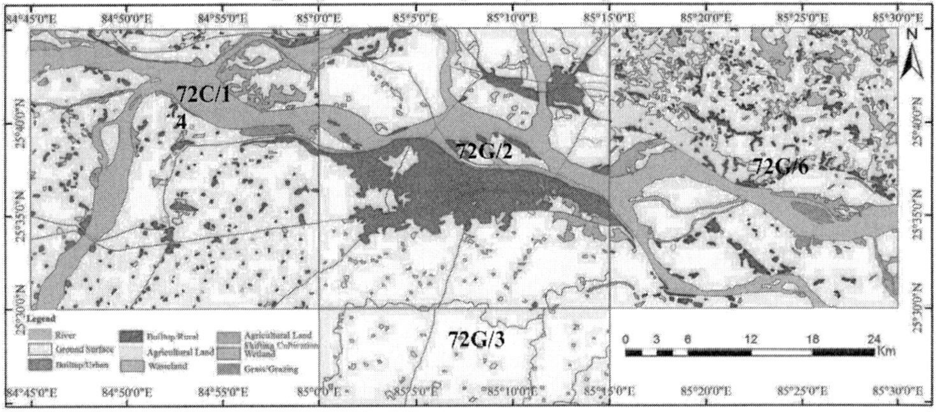

entire area of study has been collected for their analysis. This will reveal added information on the elemental constituents, productivity quotient and related parametres.

c) Land use/ land cover study: The assessment has been made with the help of satellite imageries, PGRS data ascertained by limited field checks. It has revealed the various aspects of departure from land use regulation causing environmental constraints. The assessment has also helped in identifying sectors with possibility of further growth and optimum utilization. From the Landuse and Land cover map supplemented with field checks, it is observed that the city has the potential to expand further South and South-West in toposheet 72G/2 (G/3; south of Punpun river) and towards west upto Son River in toposheet 72C/14.

d) Groundwater studies: The groundwater studies have been done to measure depth to water table along with the chemical analysis for drinking purpose evenly representing the study area. The piezometric surface varies from 24.45 m (Dharah Chak) to 8.1 m b.g.l (Taranpur) in the pre-monsoon and 14.8 m (Phulwari) to 5.5 m b.g.l. (Majhauli) in the post-monsoon period of year 2018 with fluctuations of 15.65 m (Dharah Chak) to -1.5 m (Phulwari). Assessment made in year 2019 revealed that the ground water varied from 17.5m (Gopalpur) to 7.5mb.g.l. (Rewa & Gulmahiya bagh) in pre-monsoon and 9.5 m (Gopalpur) to 2.5 m b.g.l. (Gulmahiya bagh) in post-monsoon with fluctuation range from 9.1m (Paltu Chhatni) to 2m (Korhar). The ground water contour map for 72G/ 2 shows that groundwater is deeper towards the alluvial upland in northern and north western parts and shallow ineasternand south eastern portions ofthe area.

Water Quality Index (Pre-monsoon, 2018) in parts of Patna, Bihar

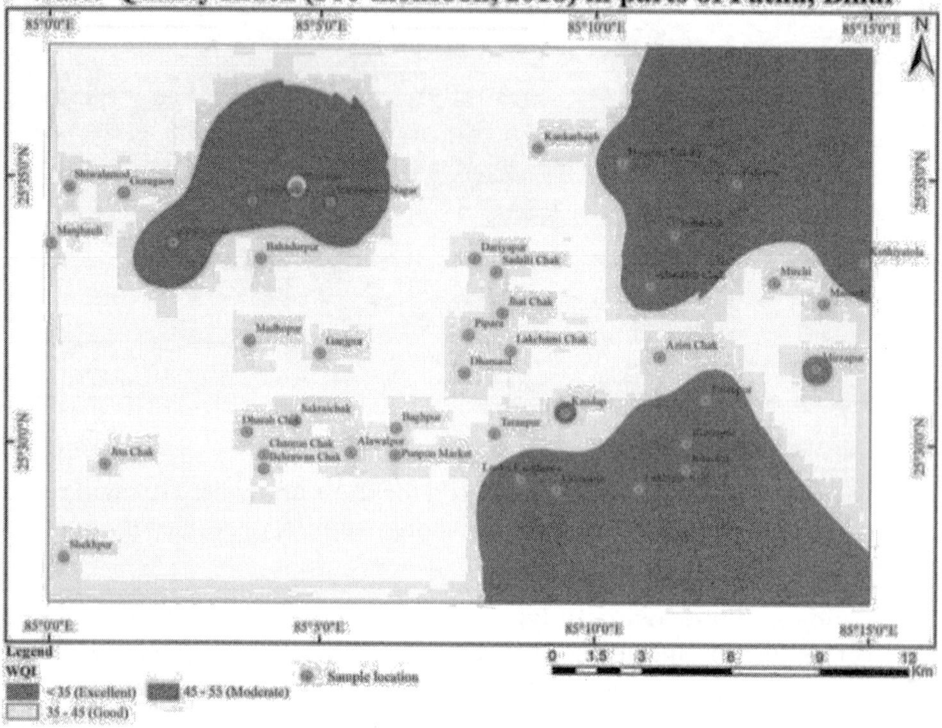

Water Quality Index (Post-monsoon, 2018) in parts of Patna, Bihar

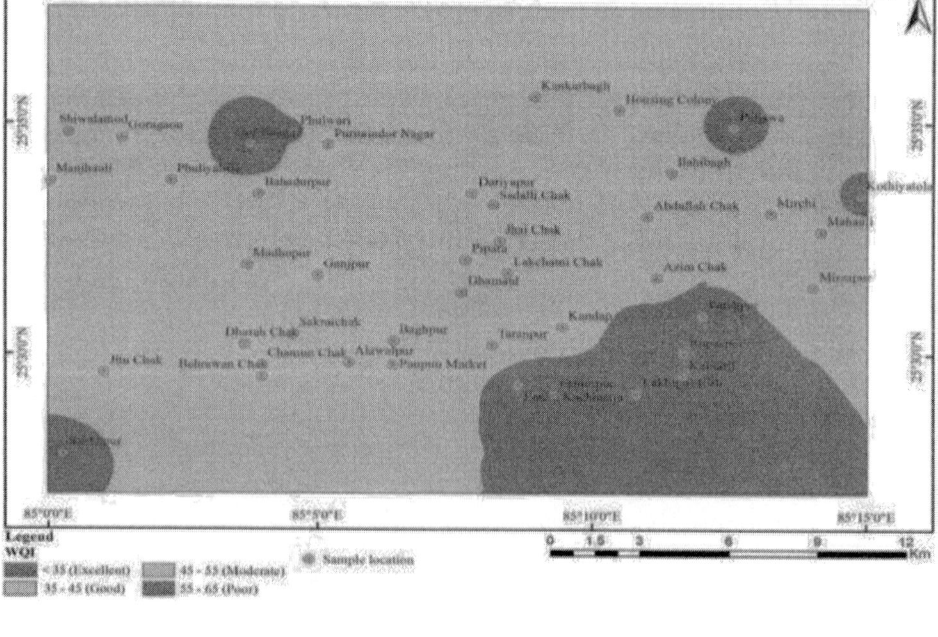

The Water Quality Index (WQI) was calculated to categorise localities with varying suitability of ground water for drinking purpose. The spatial distribution map of the WQI shows that the central part of the area (Gopalnagar, Phuliyatola, Madhopur, Dhurdhurtola, Bahtehri, and Majhauli) shows high value of Na, K, NO3, HCO3 and F. The eastern part of the study which is well urbanized shows high values of pH, TH, TDS, Ca, SO4 and Fe. The northern and southern part of the area to the southand north of River Ganga and Punpun respectively lie in the excellent category for meeting the drinking purpose of groundwater. The western part (Doghra, Bazidpur, Kanhauli) of the study area shows high value in pH, TH, TDS, Ca, Mg, NO3, Cl, SO4, F, Zn and Fe. Approximately, ~90% of study area falls under the category of 'Excellent' with regard to the drinking purposes.

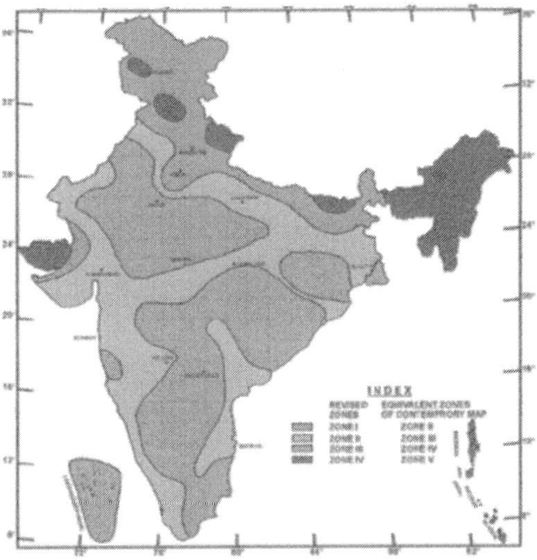

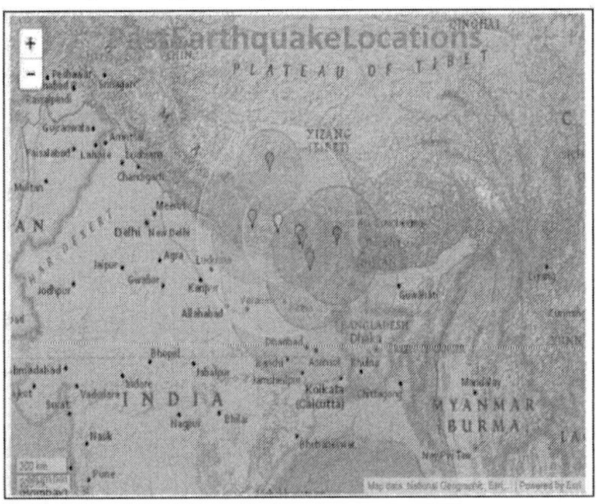

Source: IS,1893-2002

Seismic Zonations map of India

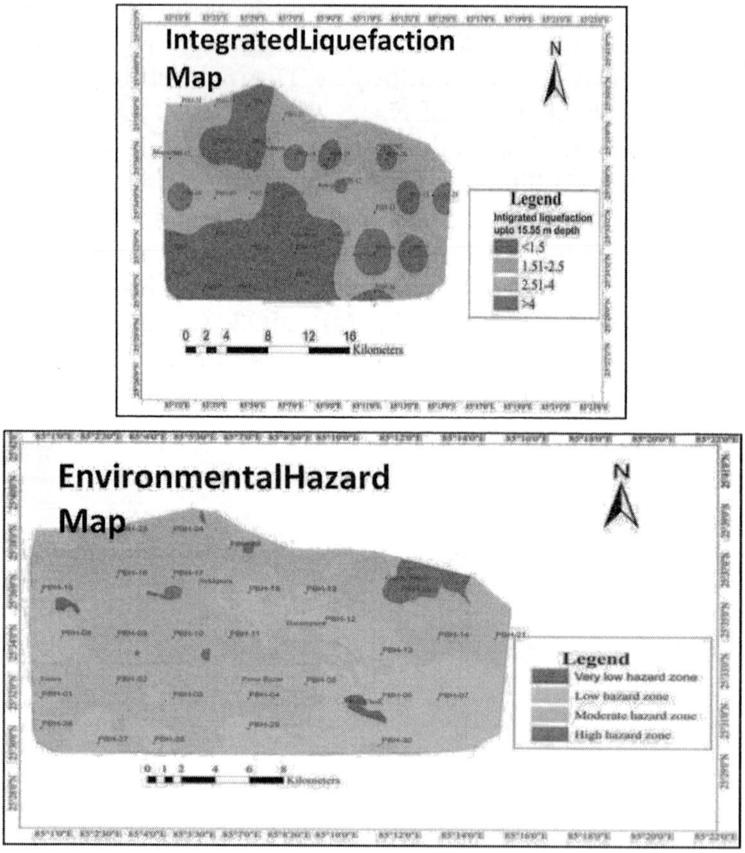

Source: IS,1893-2002

e) Seismic Vulnerability: Patna lies in Zone IV of the Seismic Zoning Map of India. Seismic Microzonation has already been attempted for toposheet no 72G/2 in FS 2012-13 & 2013-14 for anarea of 300 km^2. The findings are incorporated in the report. The geophysical studies have been carried out to derive site response signatures by noise survey and vertical electrical sounding in toposheets 72C/14, G/2 and G/6. Amplification factor varies from 1.12 to 4.43 and peak frequency (PF) ranges from 0.8 Hz to 8.27 Hz. Behrawan chak and Painapur lies under low peak frequency zone showing thick sediment cover and the same is also supported by high amplification factor. Highpeak frequency at Chinea Bela and Madarpur area indicates the relatively thin sediment thickness and low amplification factor. Are presentative resistivity profiles showing four layered structures with HKH, H, HK, H-type of curve giving idea of subsurface lithology is also attempted.

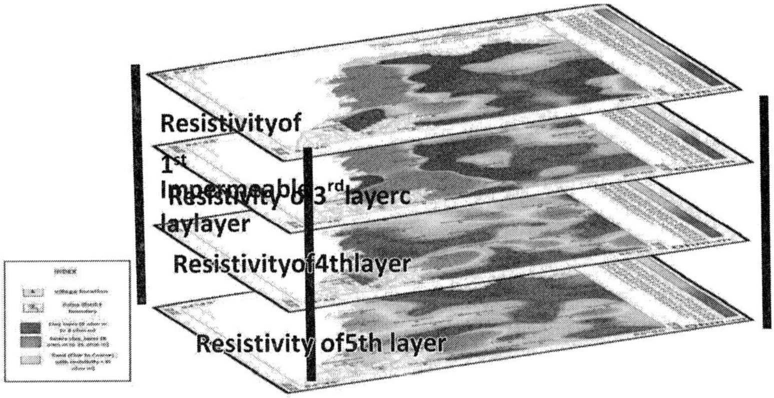

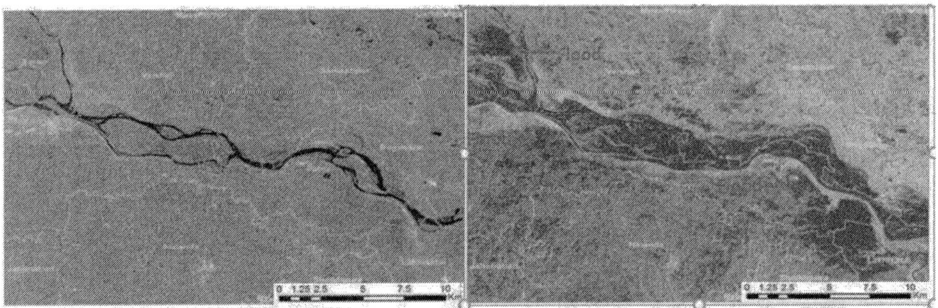

Pre data: 04th August, 2019

Post data: 30th September, 2019

Source: National Remote Sensing Centre (NRSC), 2019

Villagers walk through 1975, flood Patna, Bihar

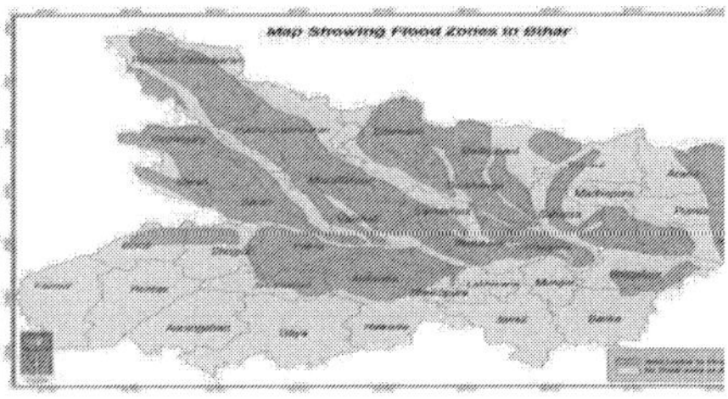

Flood Zones in Bihar(Source, BMTPC, Indi

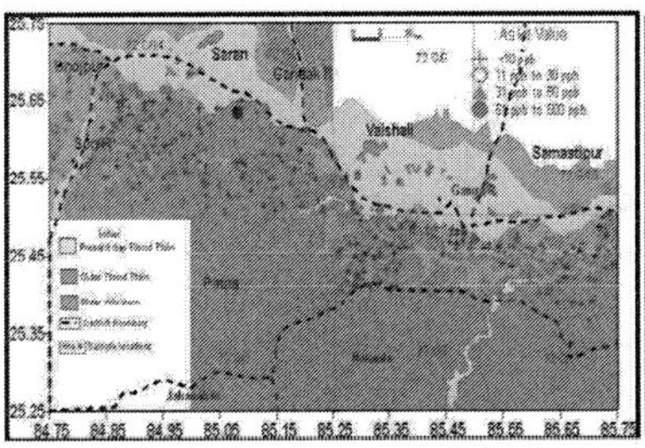

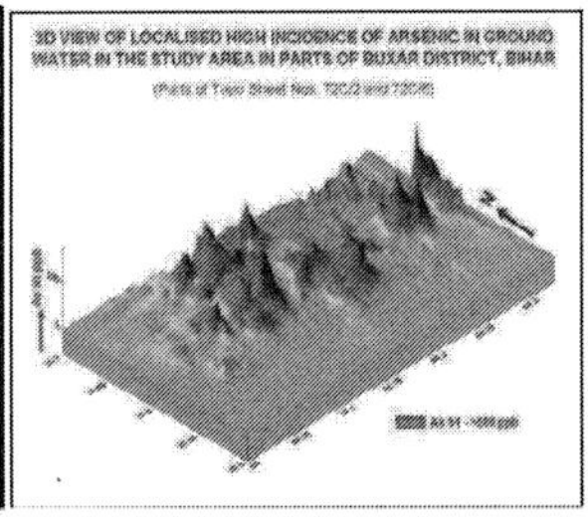

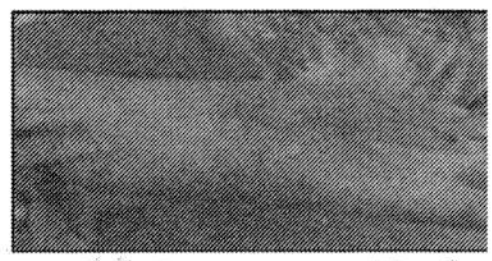

Health Impact on High

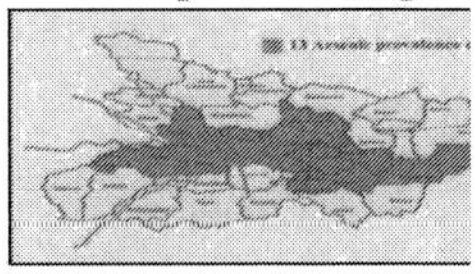

Arsenic Contamination in Patna District, bihar

Arsenic Prone Distri

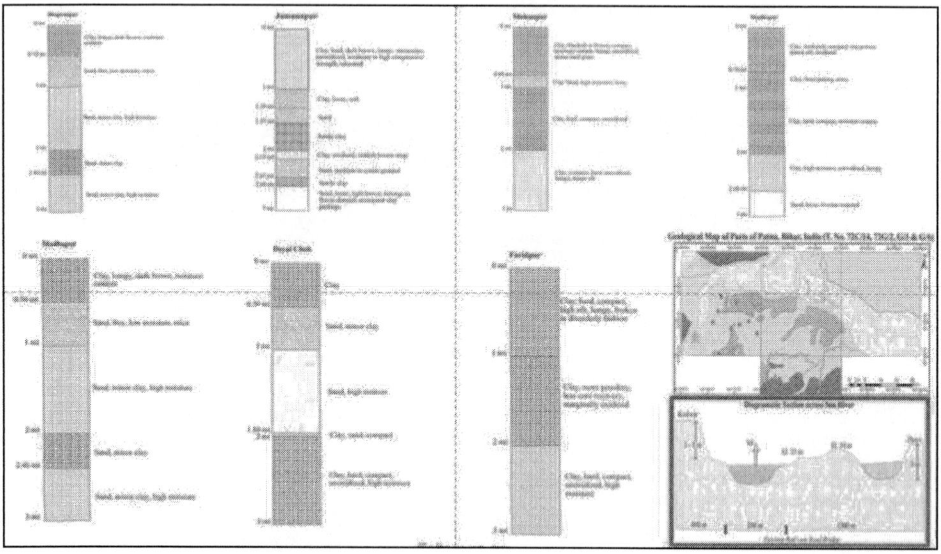

The areas in proximity to major rivers were studied in detail with emphasis on river morphology. The three rivers viz. Son, Punpun and Ganga were assessed with respect to their potential to inundate the low lying areas along their banks. There is a man-made embankment all along the Son and Ganga River which protects water from inundating into the land areas. Still there are some localities such as Sherpur (south of Ganga) where the earthen embankment over a period of time has either been damage dor is not adequate to deal with sudden flash floods leading to water ingress. The cross-sections have been prepared at many locations along all these rivers which may be helpful for flood management planning.

Problems of biogeochemical hazard of Arsenic contamination in groundwater in and around Maner within the Holocene sediments has been identified and the morpho stratigraphic and depth control established by extensive sampling and drilling.

f) Near subsurface foundation studies: Near subsurface studies have been accomplished on grid pattern down to a depth of 3m (~10 ft) from the plain table land with the help of improvised hand auguring. The geotechnical parametres will help in deriving foundation conditions for light foundation structures. The analytical results lead to following conclusions:

1. Most of the soil are fine grained, clayey/silty clayey and inorganic in nature having low to moderate compressibility.

2. Soils are in general suitable for light structure foundation.

3. For heavy structure, site specific study, deep foundation, heavy piling and other reinforcement measures may be required depending on design specifications.

4. Soil appears suitable for construction of embankments.

The Environmental Management Plan (EMP), as a product of Environmental Impact Assessment and Management is recommended which will include creation of new municipal zones in the growth indicated area. The scope for "green zone" has been duly included in the study. Land utilization foroptimum use is suggested giving cognizance to the physical, chemical and hydraulic merit of land, water and soil. Understandably, correctivemeasureshave also been suggested.

Perpetual concerns and corrective short and long term measures have also been attempted. Lack of effective sewage system, improper land use practices, ill planned communication network, unchecked human interference and unplanned and haphazard growth of urban agglomeration are identified as the contributing factors to geo-environmental problems in the area.

All this calls for a proper planning of city and modifications in light of geoenvironmental indicators, in the present landuse pattern.

References :

BIS (Bureau of Indian Standard) (2012) 10500; Indian standard drinking water specification, Second revision, pp.1-24.

BIS 189-1 (2002); Criteria for Earthquake Resistant Design of Structure.

Bose, P.K., Saikia, B. C. & Mukharjee, B. B. (1964-65); "Refraction Seismic Survey for Ground Water in Patna and Gaya District, Bihar"; Geological Survey of India, Unpublished report.

Brown RM, McCleiland NJ, Deininger RA and O'Connor MF (1972) A water quality index - Crossing the Psychological barrier (Jenkis, S.H., ed) Proc. Int. Conf. on Water Pollution Res., Jerusalem, Vol.6, 787-797.

Central Groundwater Board (2013); Groundwater Year Book; 2012-13.

Central Groundwater Board (2012); Groundwater Year Book; 2011-12.

Dayal, N., Sinha, Ravi Kr., Singh, Satyendra Kumar, 1995-96 & 1996-97; "Final report on Geo environmental appraisal of Patna Urban and surrounding areas, Patna district, Bihar"; Geological Survey of India, Unpublished report.

Horton RK (1965) An index number system for rating water quality. Journal of Water Pollution Control Federation, 37, 300-305.

Joshi, D. D., Joshi K C (2015); "Macro seismic and aftershock studies of 25 April, 2015 Nepal earthquake within Indian territory"; GSI, unpublished report.

Mangukiya R, Bhattacharya T & Chakraborty S (2012) Quality characterization of groundwater using water quality index in Surat city, Gujarat, India. International Research Journal of Environment Sciences, 1, 14-23.

WHO (1996): Guidelines for drinking water quality. Second edition, v.1, Recommendations. World Health Organization.

6

Sustainable Thinking and Applied Sustainable Development: Experiments to Responsibility

Dr. Tapas Pal
geo.tapaspal@gmail.com

Abstract:

In today's world, we are going through a situation where the most mischievous acts of human beings and the snatching of everything from nature are everyday occurrences. People today have delivered plastics to animals only for the sake of money and continue to harm them in various ways. Through the use of inadequate plastic, now peoples want to be covered with a plastic sheet on earth. Randomly throwing solid waste everywhere is causing damage to nature as well as showing irresponsibility. As a result, nature has gradually lost its balance. United Nations Sustainable Development Goals set a target for the future, we all want sustainable development, but we are not trying to make it in a reality. People are not aware of their nature and actions. That's why 'we need a climate leader' (plate and some captions to awaken awareness among the people at present. For which I have created so many captions 'We Need Climate Leaders' (plate 6), 'Stop the Religion vs. Religion, follow Works vs. Works' (plate 7), 'Nature or Environment: what you want?' (plate 4), 'Lockdown for Climate' (plate 9), 'Vote for O2', etc.

Keywords: Sustainable Thinking, Sustainable Development, UN 2020.

The basic question is raised among us that why do we want to discuss this type of differently-abled topic? Presently, we need sustainable development. We are going towards the vision and mission of the United Nations 2030 Agenda. So, we need to fulfill the 17 different goals of Sustainable Development. All we need a particular direction, but we are not doing proper planning on how to meet the target of Sustainable Development. Moreover, it is to be achieved by 2030 as per UNO resolved Agenda 2030 in the year of 2015. Being an environmentalist I have initiated the concept of sustainable thinking to find ways for applied Sustainable Development. Now the question is how we can apply this Sustainable Thinking in our daily life. A visit to the General Assembly at the United Nations of the USA is the inspiration for this idea. Because, it is from here, that is my indomitable desire and interest in Sustainable Development is manifested in my mind. A question must raise in our mind that which one is important whether 'Lab to Land or Land to Lab'? Standing on the knocking door of 2021, Sustainable Thinking has become urgent for us. We have all been using plastics from the past to the present, but now the amount of using plastics has increased, from where our biodiversity is not getting rest from it.

I noticed that biodiversity is carrying plastics i.e. the black identity of our current civilization. Before I was visiting Brazil, I had to go near Calcutta airport to get a 'Yellow Vaccine'. I stood there for a while, but suddenly I saw a crow carried with plastic instead of food in its mouth. When I went to Bhopal (Madhya Pradesh), I saw a monkey holding a plastic bottle inside a sanctuary (plate 10). An advertisement poster of Bisleri drinking water is found in our locality, where a camel is drinking water holding the plastic bottle in his mouth (Plate 11). In this way, by making a direct connection of plastic with animals and also it has attached to human behavior only for business. Presently the advertisement going in which direction, some selfish people are addicted to earning money to putting plastic in the mouths of animals. When I went to Dhaka city in Bangladesh, I saw a fruit shopkeeper selling each guava in a single plastic bag. Man is not aware of the nature around him. People want Sustainable Development, but the idea of making it into a reality has not awakened in their minds. Even when bought a sweet from the shop near to my house (Raiganj); I saw that the sweet is wrapped in plastic. Sustainable development is urgently needed in our lives now, but if it continues like this, we will be able to save the earth? If this continues, where will be our life going towards Sustainability or towards Unsustainability? The main water pipe of the Malda railway station from where the water is filled in the bathrooms of the railway bogies. I got an experience here, that is, wasting a large amount of water with the opening of the tap of that pipeline. We all know the importance of water at present, but we are not aware of water conservation. If we continue like this, how will we protect the groundwater level? Even one thing has observed, i.e. people waste huge milk for the worship of Lord Shiva during Shivaratri or sacrifice living animals for Kali Puja. It is also a waste of a large amount of milk. It may fill the stomachs of some poor children in our country. Again the tag of cruelty to an animal has followed here and I noticed a cow walking down the street with a plastic tag on-ear to provide her own identity. We usually use earrings to promote our beauty. But we are not excluding animals from our dirty game to satisfy our business mind. In this situation, how will we take civilization forward? When my house in Raiganj town is built, digging a hole just two feet in there brings out so many plastics. Solid waste material that you leave in the ground, does not mix easily with the soil, they remain in the soil for a long time. If we can't manage them properly, then we are in danger. People are not aware of the importance of infiltration, seepage, pond, and groundwater recharge. In the past, there was a pond where a skyscraper building is located in Sivhari Abashik Elaka near Dhaka University in Bangladesh (plate 14). Near Bolpur Rail station people throw plastics in ponds and the pond fills up a lot (plate 15). So many water bodies are getting lost day after day to build up flat houses, which we call Pukurchuri (closing the pond). When I was in Bolpur, I bought one person's food from a food shop near to the Jambunibus stand (Atithya Hotel), and noticed that eight plastics (plate 12) were used to serve only a single person's food. Just think that how do you save the environment? We are making urbanization and finishing groundwater simultaneously. Day by day the human beings are converting rural images into urban images. People are trying to wrap the earth in a plastic sheet. Is it possible to protect nature in this way? The people of a coastal region are never active in coastal biodiversity conservation. You won't know the importance of water, unless you have to walk miles in the heat, for a pot of water. So, thinking of sustainability is important to enjoy a sustainable future. In Bolpur, built a dam on the banks of the Kopai River and tried to be fixed the area of course of the river. The river has its own path, but it is being blocked. If it continues like this, then one day the river will want to return to its place. Are we happy with our education system? No, we are not happy in our present education system. You will understand the importance of one drop of water only when you do not have water. When I went to do RC (Refresher Course) at Jawaharlal Nehru University, New Delhi, there I noticed beautiful things on the ASC (Academic Staff College) campus. With just a few inches of space, a beautiful flowering tree has been growing, and given the opportunity to grow within the Campus, it has not been released to keep the road clean. This is an example of sustainable thinking. When I went to Washington-DC in the USA, I notice that people are developing several kinds of work based on sustainable thinking throughout their entire activities. They are trying to plant trees wherever they get a chance. Their pet dogs are walking around with them. They are not tied up with chains on their necks, leaving them open. If they defecate on the road they are cleaning them responsibly. This is their responsibility; this is a perfect example of sustainable thinking. We usually take lots of pictures of ourselves when we go out somewhere, but the people of New York are busier taking care of their pets than taking pictures of themselves there. There were a lot of pigeons in a house, the owner of the house tried to get the pigeons to send elsewhere, but the owner was stopped. And it is said that the birds have taken refuge here so it is wrong to chase them away. We need to think not just about the environment, but about all the animals in the environment that New York is teaching us. An individual has the right to get respect and space. There is a park in Manhattan where everyone needs a pet to enter. In other words, we are learning that we have to take great care of animals because every animal has rights to live. Green-city planning is an integral part of the modern sustainable urban environment. Terrorism has been converted into tourism using our sustainable thinking. After the terrorist attack on the World Trade Centre in Manhattan, a kind of fear spread in the minds of the people about that place and Aeroplane. But the advanced thinking style of the people improved sustainable thinking. Today that place has become a world peace centre. Which is attracting a lot of global tourists. Each of us should think about our workplace. The university chamber is our first home. That's why any universities in Albany put family photo frames in the teachers' chambers; which is rare in Indian culture. Sustainable Resource Management is another approach that I experienced from Jeju Island in South Korea and they are used to discard items for tree planting and beautification of the village. For example, bathtubs, bathroom pan, etc. are used for beautification by planting trees in them (plate 13). Jeju Island has a village where the boundary walls of the houses do not use bricks, sand, and cement. They used large stone to build it. Places along the Ocean Coast of Jeju Island are planted with trees by all the tourists who come to visit. Cultural Resource Management in Surabaya in Indonesia is an example of trying to plant trees on both sides of the road and in various open spaces. They are trying to preserve the Indigenous people and their culture. Eating at a restaurant in the city of Raiganj, there are some lines written about wasting water on a board above the washing sink, as a result, while people are going for washing hands people are must be thinking before wasting water. This is sustainable thinking. A Fuchka (Panipuri) seller is selling Fuchka by giving a message to his buyers that is "Save Water Otherwise 'Panipuri' will become only 'Puri' ...". In this way, we have to create sustainable thinking in people's minds through small steps. Apart from this, the festival of Halakarshan and Barshamangal in Santiniketan is the greatest example of sustainable thinking of Rabindranath Tagore.

I tried to initiate a 'Plastic-free Marriage Ceremony.' It was my initiative due to which my own wedding ceremony was celebrated without the use of any plastic. All the utensils were made of clay, instead of the menu card, the guest service menu was written on the pottery. The car used at the wedding had UNSDG written on the front. I have started the concept of 'Pitri Smrity Briksa.' We usually spend money on the day of the Shraddha (death ceremony) through various rituals on the day, along with building a cenotaph using bricks, sand, cement, etc. in the memory of the departed souls. But, I planted a tree in the memory of my father, so that I can keep that tree alive forever to remember my father's memory. Thus, in small steps, we need to awaken the minds of sustainable thinking people. You don't just have to remember the action on your mind; you have to come on the ground and try to convert the action into a reality. Although I have not any agricultural experiences due to my teaching profession, I tried to cultivate paddy in the field. My motive is to see that my students are inspired and develop interest in it themselves. It is a proper example of sustainable thinking. 'Gabeshana O Bangla Shafar' (Research & Bengal Journey) in my life was a milestone of Sustainable Thinking for the researchers. Most of the students make a home, car, bank balances, etc. by using the government scholarship (money) for research, but I have tried to utilize the total amount of the fellowship money for my Ph.D. research. I have traveled around 37 days across West Bengal on a motorbike to collect the data for my Ph.D. Research and cover 6631 kilometers on the motorbike. My other thought is the 'Briksha Rakhi Bandhan' (plate 5). Rakhi Bandhan festival was first initiated by Rabindranath Tagore for making a strong brotherhood relationship between the Hindu and Muslim communities. On the day of Rakhi Bandhan, I started to tie Rakhi to a tree for bonding a strong brotherhood relationship between man and tree. Because we protect ourselves, we can't accept their loss at any cost. So I am trying to protect the plant by using that thinking. 'Green Village Project' is my other brainchild with Jagori Natya Group of Raiganj. Again the movement on the caption of 'We Need Climate Leaders' started and the time has come for us to think about climate leaders, to whom we will bring to power in the present world. Then we empower those who will provide us with a pollution-free environment, freshwater, and oxygen. That is why we have launched the tree planting program in the villages in collaboration with Jagori Natya Theater. That way the greenery grows more in the rural areas. When guests come to my house, we all eat and enjoy together, and we give a sapling as a gift to the guests with the caption of 'Be Green

Project' (plate 3). In other words, through this, eco-friendly thinking and dissemination with the guest service take place among the guests. During the celebration of World Indigenous day (9th August) 2020, I celebrated it in a different way. I welcomed the Indigenous Tribal Policeman with Panchabhuta (Kshiti, Oop, Tej, Marut, and Byom). I took five kinds of fruits (make them Panchabhuta) and give them as gifts to maintain sustainable thinking. According to Hindu Holy Books, Devraj Indra's garden was called Nandankanan. Kadambini Devi makes a garden by her hand on an urban roof. The people of the city are in dire need of that gardening today. The roof of the house is usually used for drying clothes, but that roof should be used not only for clothes, but we could also use it for gardening. I started this project on my first wedding anniversary by creating a garden on the roof of my house and named it the 'Nandankanan Project' (Plate 2).

UN-2030 vision has decided a vista to achieve Sustainable Earth by the year 2030. Everyone wants Sustainable Development but we forget the actual stewardship is to nourish Sustainable Thinking. Everything is generated from our mind and after that, the thinking of the mind is converted into works, i.e. mind to work or lab to land is important. The mind can think anything like the fruitful mind for sustainable earth or destructive mind for blasting of the atom bomb on civilization. Thus, the nature of the seed of the plant provides the nature of the fruits. The basic philosophy of Sustainable Development has emerged from Sustainable Thinking. If our thinking is not sustainable how we can apply sustainable development. Every work of you and your surroundings create sustainable thinking (Plate 1) and that becomes the basis of applied Sustainable Development.

Bibliography:

Pal, Tapas. (2015). Sustainable Development. Yadava Publications Private Limited. Tiruvannamalai. ISBN: 978-81-931808-5-3.

Photos:

Contributions being an environmentally responsible person in my local area to emerge the concept of green contributions through innovative applications.

Plate 1

Plate 2

Plate 3

Plate 4

Plate 5

Plate 6

Plate 7 - 8 - 9

Plate 10

Plate 11

Plate 12

Plate 13

Plate 14

Plate 15

7

Pandemic and Exclusions: Towards A Geographical Epistemology

R.B. Bhagat
rbbhagat@iipsindia.ac.in

1. The pandemic of COVID-19 is not simply a health crisis but also triggered a huge economic and social impacts as well. The poor countries and people from lower social and economic strata were disproportionately affected being vulnerable due to the loss of lives and livelihood. The existing social and economic inequalities not only became visible but also got deepened. Therefore, it is argued that inequality is a central geographical concern in the wake of lockdown following the out-break of COVID-19.

2. The recent urban exodus of migrants from big cities of India has brought the geographical issues at the centre of social inquiry from both theoretical and epistemological perspectives. The epistemological perspective raises a critical question of how to study a phenomenon from an appropriate methodological perspective. The pandemic has also exposed the conventional approach to a problem which is largely devoid of space and place in social research.

3. Although geography has emerged as a very important dimension during the pandemic of COVID-19, the traditional geographical approach has been emphasizing the forms rather than process. As such the existing geographical methodology is highly inadequate in understanding the multi-dimensional problem unleashed by the pandemic. I would like to argue that space and place can play a central methodological tool in understanding this multifaceted crisis and also provide a plausible framework for integrating social,

economic and political processes in a spatial framework. This is likely to justify geography as an integrating discipline or geography a modern synthesis incorporating space and place and space and time as fundamental elements of geographical methodology. Geography is not simply about studying spatial variations or trying to find out some morphological regularity, but to envision spatial process at par with socio-economic and political process, and to claim geography in a way no less in importance than any other social science discipline.

4. The pandemic should be understood as a socio-spatial process because origin, spread and mitigation, all of them, reflect socio-spatial process. Lockdown, social distancing, quarantine, and containment zones are geographical categories that reshaped space and place. When we see changes in the space and place, we have to see how changes at one place affect the changes at other places. In other words, we have to look development as a spatial process. Development is not simply an economic process but spatial, ecological, economic, social and cultural as well. Mobility is a manifestation of spatial process. It shows the relationship between space and place and how it changes. The mobility is not only confined to migration but also includes flow of goods and services, capital and information. More mobility means more development and it works in an accelerating mode. It is like running on a tread mill leading to a mobility crash.

5. Geographers use the tool of space, place and region to understand the economic and social changes. We must make a distinction between absolute geography vs relative geography. This requires an alternative epistemology which can integrate space and place with economic and social changes. When our concepts and tools are not static but dynamic then it would be possible to capture the changing reality. So the question is what do we mean by the concept of being relative and relational?

6. Space is an unbounded extent defined by the location of objects and their relationship. On the other hand, when the objects are human beings it becomes a place. Place has a history, culture and social relations. There is also a contradiction between space and place. Spaces are produced, but places are made. In producing space, space becomes resources, means of production, even wealth. It is controlled and appropriated by the economically and politically powerful class. It is very much evident in the process of urbanisation creating wealth through real estate development and commodification of space. This

creates inequality and exclusions, and consequently raises a voice for the right to the city or right to the space. Space raises the question of 'where' intimately related to who gets what and where. Where we live decides about access to resources, income and employment, access to education, health care and finance. The question of where is very powerful dimension determining chances of life both interms of access to good or bad. Region is about drawing boundaries; it is about scale of space and place. Region is more related to form than to the process, although formation of region is a significant historical and spatial process.

7. When we want to understand the structure of space, I think there are three important attributes namely- location, distance, and relationship. Location can be determined by geographical co-ordinates like longitude and latitude with reference to zero-degree meridian passing through Greenwich and zero-degree latitude assigned to Equator. These references are historically determined. Further, location is not only geographic but also social, economic and political as well showing the respective locations in the social, economic and political hierarchy and structure. Same is also true for distance as well. The distance could be geographic distance, social distance or political distance in human space. The location and distance among phenomena also constitute relationships reflected namely in the spatial, economic, social and political relations. The true nature of location and distance indicates that neither of them are absolute but relative and relational in character. Further the relationship is not static but dynamic shaped by various flows namely flow of goods and services, capital, people (migration) and information on the geographic and human space.

8. Geographically, relationship manifests in various forms. One of the forms of relationship is density. The density is no tonly about population but also about economic activities. In fact, when population and economic activities are concentrated in a unit of space taking the form of urbanisation. Ironically, our urbanisation is extremely exclusionary dominated by big cities. It is neither promoting balanced regional development nor environmentally sustainable. The exclusionary nature of urbanisation has been the driving force widening inequality supported by economic and urban policies based on neo-liberalism and globalisation. Neo-liberalism takes the form of privatisation, formalisation, legalisation, digitalisation etc. on the assumption that the economic growth will trickle down. Economists supporting neo-liberalism and globalisation

argue for more reforms for economic growth to be undertaken by the state, but the empirical reality continues to show polarised economic development and increasing inequality at various spatial scales. Inequality limits the human ability of freedom (freedom from hunger, freedom from disease and freedom from exploitation) and the idea of fraternity i.e. sense of belonging and inclusion very evident during the exodus of inter-state migrants. A question is being raised whether the migrants belonged to the state of origin, or state of destination or they are the responsibility of the central government.

9. Since we want to see how the present conditions change. We have to see how space is also related to time. There is a contradiction between space and time. Time is related to future outcomes shaping our conditions, but we often do not give equal importance to space shaping our time. Therefore, there is disjoint between the two. We often see that the political parties make promises to make our life better. It is a type of politics of time. However, we have to also examine the politics of space behind such politics of time. Therefore, planning, decentralisation and development must be looked from space and place point of view. The promise of Achche Din (politics of time) must be accompanied by appropriate politics of space in favour of planning over project, decentralisation over centralisation, and local over global. In this context the long standing demand of special package for the development of Bihar assumes significance for achieving regionally balanced development. Moreover, the causes and consequences are not merely the empirical reality, but for a deeper clarity we must look into the problem from methodological and epistemological perspectives as well. The pandemic has unequivocally demonstrated the importance of geographical epistemology in understanding its origin, spread and mitigation from the lens of space and place.

References :

Abler, R., Adams, J. S. and Gould, P. (1971) **Spatial Organization**, Prentice-Hal, Englewood Cliffs.

Bhagat, R.B. (2006) "Space and Social Research: Some Methodological Issues", **The Indian Journal of Social Work**, Vol. 67, No. 1 & 2, 2006 pp. 82-91.

Bhagat, R.B., Reshmi, R.S., Sahoo, H. Roy, A. and Govil, D. (2020) "The COVID-19, Migration and Livelihood in India: Challenges and Policy Issues: Challenges and Policy Issues. **Migration Letters**, 17(5), 705-

718, 2020, https://doi.org/10.33182/ml.v17i5.1048.

Dorren, Massey (2005) **For Space**, Sage Publication, London.

Harvey, David (1990) "Between Space and Time: Reflections on the Geographical Imagination", **Annals of the Association of American Geographers**, Vol. 80, No. 3 (Sep., 1990), pp. 418-434.

Mazúr, E.(1983) "Space in Geography" **GeoJournal**, Vol. 7, No. 2, pp. 139-143.

Scargill, D.I. (1985) "Space, Place and Region: Towards a Transformed Regional Geography", **Geography**, Vol. 70, No. 2, pp. 138-141. Smith, Susan J. (1989)

"Society, Space and Citizenship: A Human Geography for the 'New Times"? **Transactions of the Institute of British Geographers**, Vol. 14, No. 2, pp. 144-156.

8

Groundwater Arsenic Poisoning and Health Status of the Exposed Population of Bihar

Arun Kumar and Ashok Kumar Ghosh
drarunk31@gmail.com

Introduction

In the present times, arsenic catastrophe has become a major challenge for the exposed population. It has become a global problem and is estimated that >300 million persons are exposed to arsenic by drinking arsenic contaminated water. In India it is estimated that about 70 million persons are exposed to arsenic poisoning by consuming arsenic contaminated water. Moreover, the Ganga- Meghna-Brahmaputra plains are thought to be the area where the arsenic catastrophe has recently raised many folds. Bihar is a landlocked state which lies in the lower middle Ganga plain and the state is also drained by seven major tributaries, originating from The Great Himalayas. Since pre-historic times, due to the fertile land area of the plains of the river Ganga, civilisations flourished along the banks of the holy river. Hence, the region is thickly populated. Arsenic menace in these areas by poisoning the groundwater has become a major challenge as serious health related issues are reported. In the state of Bihar, an estimated 10 million people are exposed to arsenic. Groundwater is the primary source of arsenic poisoning which has caused health hazards to the exposed population. The exposed population are exhibiting typical symptoms of arsenicosis such as hyperkeratosis, melanosis, pigmentations and other skin manifestations. Apart from this they are also exhibiting other symptoms such as disorders of liver, kidney, nervous system, cardiovascular, hormonal etc. The arsenic exposure firstly, makes the exposed population diseased and if not cured are at the risk of getting

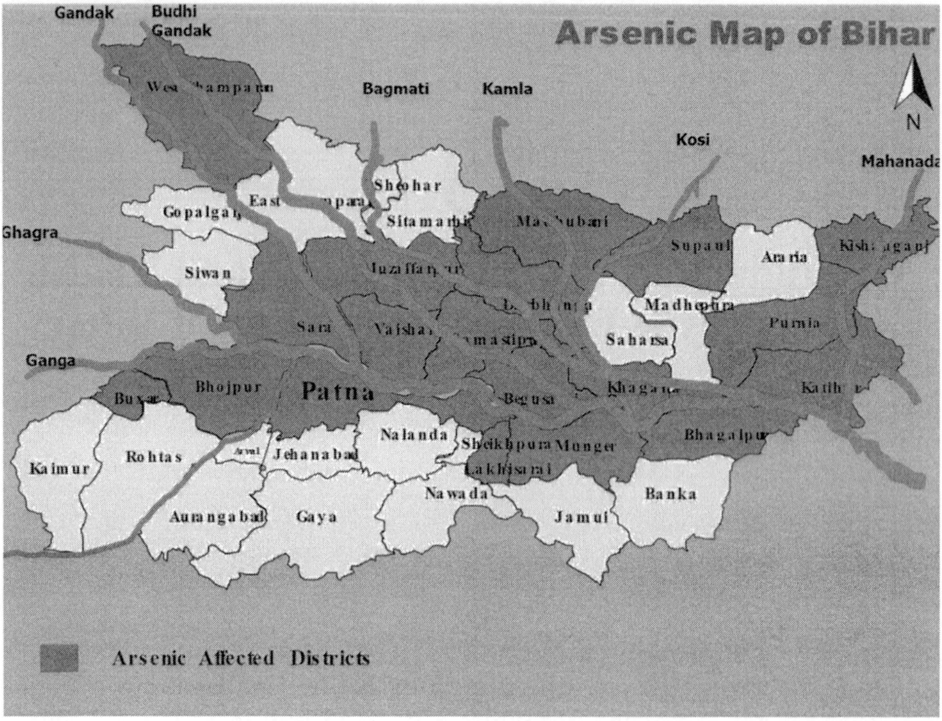

Fig. 1. Arsenic Map of Bihar- 2019 (Districts having arsenic contamination in ground.

the disease such as cancer. The cancer among the exposed population is majorly of skin, liver kidney, gall bladder, bladder, breast, colorectal etc. Hence, it is very important to evaluate the health problems exhibited in the exposed population.

We have evaluated the various geological terrains of Bihar and have found that the arsenic poisoning is very common in the river basins of north Bihar as well as the Gangetic plains. This north Bihar arsenic affected river basin districts are Buxar, Bhojpur, Patna, Lakhisarai, Munger, Bhagalpur, West Champaran, Saran, Muzaffarpur, Vaishali, Samastipur, Madhubani, Supaul, Darbhanga, Begusarai, Khagaria, Kishanganj, Purnea and Katihar (Fig.1).

Health Status of the Exposed Population:

The arsenic exposed population exhibit typical skin manifestations such as keratosis, melanosis, leucomelanosis, rain drop pigmentations and disease of cancer such as skin cancer, lung cancer, renal cancer, bladder cancer, gall bladder cancer, liver cancer, colo-rectal cancer etc (Fig.2).

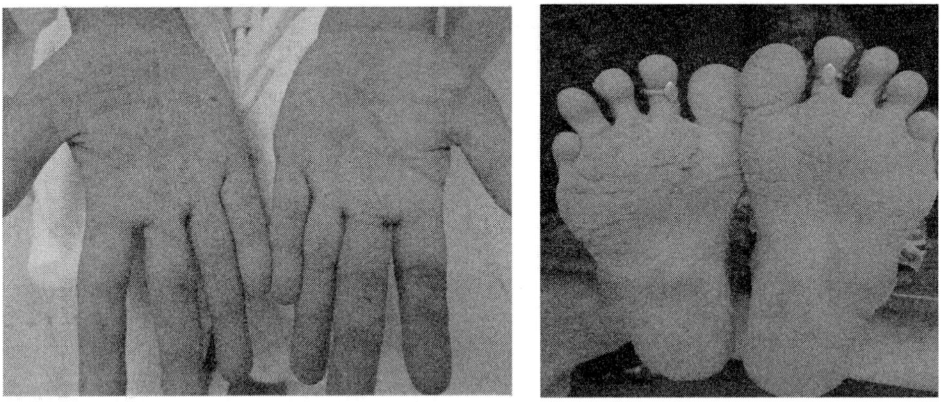

Fig. 2. Arsenicosis symptoms in exposed population

Altogether, the arsenic exposure weakens the body of the subjects, causing disease in them and if the same disease is either ignored or not cured generally leads to cause disease of cancer. The mode of arsenic poisoning to human health mainly depends upon the chemical forms of arsenic such as As3+ or As5+. Both the forms are highly toxic but especially the trivalent form causes major health hazards. The trivalent arsenic usually enters the human body through drinking water and is absorbed in to the blood and is transported to the vital organs of the body especially, the liver and kidney. The liver however reduces the toxicity by converting it furthermore to the less toxic compound Dimethyl Arsenic Acid (DMA) and finally eliminating it by the kidney. However, the DMA is also a carcinogen which if remains in the system, causes toxic effect to the vital organs. It furthermore also hampers the normal hormonal functioning as it disrupts the functions and acts as xenoestrogen. It influences the functions of the other hormones such as thyroid hormone function, PPAR receptor function, testosterone, progesterone and estrogen receptor functions. Apart from this it also disrupts the functions of glucocorticoid, mineralocorticoid and retinoic acid receptor functions. Hence, arsenic exposure causes serious assault at the molecular level in the human body.

Recommendations

Mitigation interventions are immediately required in the exposed area of Bihar, since the exposed population's health is deteriorating day by day. Following are the intervention methods which can be implemented -

1) Use of surface water.

2) Use of rain water through rain water harvesting.

3) Use of ground water with acceptable level of arsenic.

4) Use of arsenic filter units.

Moreover, for the exposed population, following health strategies needs to be implemented -

1) Regular health check-ups for the exposed population by organising health camps.

2) Regular monitoring of the source of drinking water of the exposed population.

3) Motivation to the exposed population for the usage of arsenic free water.

4) Awareness among the exposed population related to their health.

9

Indoor Air Pollution: The Invisible and Silent Killer

Saleha Jamal
salehajm@gmail.com

From smog hanging over cities to smoke inside the home, air pollution poses a major threat to health and climate. While the battle to check ambient air pollution has been in the news for a long time, the less-researched hidden air pollution in homes, offices and other indoor environments is no less dangerous for health. Not only are ambient air pollution and indoor air pollution inextricably linked, globally, people spend about 90% of their time indoors. Indoor air pollution is an issue that impacts a wide range of enclosed spaces such as houses, offices, schools and public transports, in which we spend 80% of our time on average. Whether caused by fuels used in the home or the presence of toxic materials, chemicals and pollutants in the ambient air, the effects on health can be extremely severe. Human-induced in door air pollution is to a large extent caused by household cooking fuel used for cooking and heating, usually involving open fires or traditional stoves in conditions of low combustion efficiency and poor ventilation. As aresult, there are high levels of air pollution to which women, especially those responsible for cooking are most heavily exposed. During cooking, the pm_{10} concentration in an Indian kitchen varies between 500-2,000 ugm^3 and as women spend 3-7 hours in the kitchen every day thus, she is likely to breathe in smoke equivalent to consuming two packs of cigarettes.

Indoor air pollution also originates from other modern indoor air pollutants associated with industrialization, with a variety of suspected health effects such as sick-building syndrome. However, from a global burden of disease point of view, these modern indoor air pollutants are relatively minor (Larsen et al., 2008).

Few statistical facts about Indoor Air Pollution/household air pollution:

- 1.2 million deaths per year from exposure to household air pollution
- 100,000 child deaths per year from exposure to household air pollution (WHO, NFHS, 2016)
- Some 1.7 million Indians died due to air pollution in 2019 (report by Lancet Planetary Health).
- Air pollution exposure, 476,000 infants died in their first month of life in 2019 (globally).
- 116,000 infants in India died within a month of birth in 2019 due to air pollution - outdoor and indoor, of which household air pollution (from burning solid fuels for cooking) accounted for about 64 per cent of all neo-natal deaths attributable to air pollution. Infant deaths in India were followed by those in Nigeria (67,900), Pakistan (56,500) and Ethiopia (22,900) (Down to Earth, 2020)

Thus, it can be said that ***Indoor air quality is an often-forgotten issue in the field of air quality***. Indoor air pollution and urban air quality are listed as two of the world's pollution problems (World's worst polluted places report, 2008) but indoor air pollution which is generally less understood and neglected accounts for twice the number of deaths reported from urban outdoor air pollution (Down to Earth, 2007).

According to published earlier this year by the Boston-based Health Effects Institute, an independent global health and air pollution research organization, an estimated 846 million people in India were exposed to household air pollution in 2017 (the State of Global Air Report, 2019).

Indoor air pollution is the most direct physical health risk. While the precise mechanism of how exposure causes disease is still unclear, it is known that small particles and several of other pollutants contained in indoor smoke cause inflammation of airways and lungs and impairs the immune response. High concentration of indoor air pollution increases the risk of acute respiratory infections (Acute Respiratory Infections) in young children, chronic lung diseases and cancer in adults and adverse pregnancy outcomes (such as still births) in women exposed during pregnancy. On the basis of site of infection it is referred to as ARI of upper (AURI includes common cold, pharyngitis and otitis media) or lower (ALRI includes epiglottis, laryngitis, laryngotracheitis, bronchitis, bronchiolitis and pneumonia) respiratory tract (Park, 2007).

How the toxic pollutants effect human health/kills

- The PM10 levels in homes using biomass range from 300-3,000 μgm3. This can also go upto 10,000 μgm3. During cooking the PM10 concentration in an Indian kitchen varies between 500-2,000 μgm3.
- Inefficient burning of solid fuels indoor emit a mist of pollutants including carbon monoxide, particulate matters, nitrogen oxides, benzene and polyaromatic hydrocarbons. The fine particles PM2.5-less than 2.5 micrometers in diameter penetrates into the respiratory system.
- Women spend 3-7 hours in the kitchen everyday and breathe in smoke equivalent to consuming 2 packs of cigarette. This causes acute and chronic respiratory cardiovascular diseases.
- Pollution also causes pneumonia and other acute lower respiratory infections (ALRI) among children and chronic obstructive pulmonary diseases (COPD) and lung cancer in adults.

Keeping these aspects in mind **the focus of my work is on the health problems occurring due to indoor air pollution from type of household energy used in different income households of Aligarh city**.

Data base and Methodology

The study is mainly based on primary sources of data which were collected through household surveys from rural and urban areas with the help of questionnaire interviews. The following methods were adopted for the study (Jamal, 2012),

i. For the purpose of selecting the sample, hundred households belonging to different income strata (High, Medium, Low and Very low income group) were sampled. The total sample size consisted of 400 households. From each household woman respondents actively engaged in household/cooking work were chosen. Information regarding the socioeconomic characteristics, domestic fuels used for cooking and indoor air quality were collected.

ii. Indoor air pollutants were monitored in 5 per cent households (20 households) of the total sample (400) in different cooking places (in separate kitchen with ventilation, separate kitchen without ventilation, in multipurpose room, in verandah and in open air) using different domestic cooking fuels (traditional/biomass fuels, modern/

LPG) to diagnose the indoor air quality. For the purpose of monitoring of SPM (PM10, PM2.5), a handy sampler "Portable GRIMM Dust Monitor Series 1.109" and for monitoring gaseous pollutants (CO2 , CO, SO2 , NO, NO2) "YES-206" and "YES-205" handy samplers were used.

iii. Data regarding the health conditions associated to indoor air pollution were collected through (i) personal interviews with the help of the questionnaire and (ii) from the records of the Out Patient Departments of Jawahar Lal Nehru Medical College Hospital, Malkhan Singh Government Hospital, Mohan Lal Gautam Women's Government Hospital and various private doctors' clinics.

iv. Risk factors associated to indoor air pollution were identified and Karl Pearson's correlation co-efficient (r) method was used to examine the relationship between indoor air pollution related risk factors and associated diseases.

Findings

- Even today, when most people think of air pollution, they think of outdoor air pollution and not indoor air pollution. The source of pollution with greatest health consequences remains unseen.
- In Aligarh the use of traditional fuels and inefficient stoves remains to be the most important source of indoor air pollution in many of the poor (low income) households.
- Nearly all the low-income households were dependent on traditional fuels while few of the medium and low-income households were using a combination of both fuels (traditional and modern)
- Women are the main sufferers to indoor air pollution
- Concentration of pollutants was very much high during cooking hours as compared to non-cooking hours and the concentration of pollutant emissions from biomass fuel was very much higher compared to the emissions from LPG.
- Low-income households were most prone to indoor air pollution because they were living in sub-standard houses, having just one room used for cooking, living and sleeping; characterized by no ventilation, congestion/crowding; food being cooked on traditional fuels/stoves and spending long hours in kitchen.
- Mostly the low income women/households bear the burden of associated diseases (ALRI, AURI, COPD, asthma, pulmonary

tuberculosis, perinatal mortality, low birth weight, cataract and eye irritation) mainly caused by the use of traditional fuels.

In general, use of traditional fuels and stoves are the outcome of low income of households which leads to rigorous exposure to the emitted toxic pollutants leading to the increase of risk of a number of problems and diseases like cough, phlegm, eye irritation, acute respiratory infections (ALRI, AURI), chronic pulmonary diseases, asthma, pulmonary tuberculosis, perinatal mortality, low birth weight, cataract etc.

Finding solutions to indoor air pollution

- Need to recognize the vulnerable households/areas affected due to indoor air pollution
- Various agencies including Ministries of the States and Central government, Planning Commission, research institutes, non-governmental organisations (NGOs), multinational organisations and local Municipal government should develop strategies to reduce indoor air pollution. Regular monitoring of the implemented plans is the need of the hour.
- Interventions can occur at many fronts, as like:
  - improving cooking devices (fuels, stoves, cooking appliances),
  - improving living environment (ventilation through doors and windows, separate kitchen) and
  - behavioral change (spending less time in kitchen, fuel drying, using pot lids,
- Certain suitable strategies should be adopted, as like:
  - Choice of clean and efficient cooking fuel (LPG, electricity, etc)
  - Use of improved stoves providing more heat, less smoke and less fuel consumption
  - Development of cheap and less polluting energy sources like biogas, solar heaters etc.
  - Provision of proper ventilation condition, both in the cooking area and the living area
  - Awareness towards the ill effects of indoor air pollution.

References:

State of Global Air Report 2019, published earlier this year by the Boston-based Health Effects Institute, an independent global health and air pollution research organization,

Down to Earth 2007. July 15, Centre for Science and Environment, New Delhi.

Down to Earth 2020. October 21, Centre for Science and Environment, New Delhi.

World's Worst Polluted Places Report, 2008, Blacksmith Institute of New York in collaboration with Green Cross Switzerland.

Park K., 2007, Prevention and Social Medicine, Banarsidas Bhanot Publisher, Jabalpur, India

10

Re-Imagining the World Cartographically

Sarfaraz Alam
sarfarazalam05@gmail.com

I like maps, because they lie.
Because they give no access to the vicious truth.
Because great-heartedly, good-naturedly
they spread before me a world
not of this world.
–Wislawa Szymborska (translated from Polish by Clare Cavanagh)

Introduction

Humans have always struggled to comprehend and make sense of their vast and complex home - the earth. And their struggle continues in some form or other. To overcome this struggle, geographers and cartographers have devised a 'dependable tool' in the form of a map that provides information about the earth, such as distance, direction, location, height, size, shape, etc. As a scaled-down model, maps represent the earth or its parts in varying details. They help human beings through a simplified view of the earth, a picture, which is, to a large extent, reliable or at least thought to be so. These qualities make maps an integral part of our everyday lives whether drawn on a piece of paper or through their existence in our mind. One can comprehend this because we often think about places either through paper maps or mental maps. Such a conviction can indeed be related to the fact that they shape our understanding of the places and also help in the creation of a perception about them.

However, maps are not always the accurate and the most objective representation of the earth or its parts. For example, the transfer of a three-dimensional world into two-dimensional figure leads to distortion of some characteristics of the earth, its shape, size, direction, or orientation. Further, since maps represent the world to make it widely understood, they impose a structure on the world.

All communication depends upon accepted conventions. If there is no agreement, there is no mutual understanding. Because it is a conventionalized form, the map imposes a structure on the world, rather than merely describing it. Thus maps, like all other graphic forms, are not neutral. How something is depicted - or even suggested - is as strongly meaningful as the subject itself. A map imposes a restricted view of the world for a particular purpose. To use a map correctly, its specific structures must be learnt (Treib, Marc, 1980: 6).

In some cases, maps are deliberately made to exaggerate the size of different places on the earth. It is due to the human tendency to unconsciously equatingthe extent of territory and its location with national prestige and power. In other words, the size of territory a country occupies is correlated with its power and its accessibility to natural resources. Hence, there is a tendency to exaggerate the size of territory through maps, the most common example being the use of Mercator projection that shows the countries located in higher latitudes disproportionately that is more extensive than their actual size. Similarly, many people consider their country as the centreof the earth. While many believers of Islam consider Mecca as the centre of the world, and for manybelievers of Hinduism, it is the Ujjain city. The British drewthe Prime Meridian on their world map to pass through the Greenwich Village (near London) after an international agreement in 1884. It is the line of zero degree longitude and therefore the starting point of measuring distances of all the placeslocated to its east as well as west. Since Prime Meridian is an arbitrary line, it can be chosen to be anywhere (National Geographic, 2021). The French and the Chinese did not agree with the idea of Prime Meridian passing through Britain. They wanted the Prime Meridian to cross their own countries in world maps.Therefore, their cartographers published maps with zero degree line passing through Paris and China, respectively. Thus, the final decision of the country from where the Prime Meridian would crosswas decided politically in 1884.

Given these, anyone reading a map but not aware of its technical limitations will develop false as well as partial ideas of the earth or its parts. Therefore, a certain amount of exposure and education is required to understand any map (Treib, Marc, 1980: 15). Unfortunately, not all of us have the education and exposure to understand any map correctly.

Since maps are simplified representations of the earth and its places, they can not show all the spatial elements. It is true even for representing smaller places, such as urban and rural settlements. There is a tendency among cartographers to show some parts of a city while ignoring 'the others'. As a result, the spatial elements of a city are over-represented (magnified) and under-represented (understated) or not represented at all. For instance, the tourist map of a city does not give equal importance to all the localities. It shows thelocalities of tourists' destinations with bright colours, while other places with faint colours. Thus, the whole mapping exercise is subjective and selective. What is more distressing is that they altogether avoid representing 'dark places' (e.g. brothels, squatters, dirt drains, etc.) on a city map. Marc Treib (1980: 19) believes that as a selective tool, tourist maps of cities represent only those spatial elements that are positive and desirable rather than those designated as negative and undesirable. As a result, we never see slums, squatters, buildings in poor conditions, in tourist maps. But the reality is that tourists develop images of the cities through these tourist maps and carry with them the cities' images that arenot only incomplete but also biased.

Maps present the world but, they do not always tell the truth. In other words, maps lie and that too in a deceptive manner; that the map reader believes whatever maps communicate isaccurate and truthful. But in reality, maps deny access to the truth about places by representing them in discriminatory manner. The selections are not always objective. The inclusion and exclusion of spatial elements in a map are deliberate and biased, usually favouring the ruling elite. A city map, for instance, is about the city only to the extent the ruling elite want it to be. Maps therefore, are subjective as well as ideologicalserving the interests of those in power.

But this is just one side of the story. If map-making is truly selective, then maps can be constructed to highlight the worlds which otherwise remain concealed and excluded from the mainstream. Maps can also visualize alternative worlds, i.e., 'the worlds' that are either not represented or under-represented through conventional maps (e.g., 'negative' and 'undesirable' worlds). Similarly, maps canmake us imagine the world differently, and can form and mobilize opinions on specific issues and bring social transformation. They can be a powerful tool in the hands of the powerless to re-imagine the world and reclaim 'their worlds' cartographically. Therefore one can say that for the subaltern communities, alternative mapping is a starting step for physically reclaiming their places in the world.

This chapter examines the need to bring in the concepts of society and space in the centre of mapping practices in the backdrop of this discussion. Fortunately, in recent decades, mapping practices are changing for good. New maps are about 'alternative ways of visualizing the world' and 'visualizing of alternative worlds.'

Society and space

Space and society are widely considered the fundamental concepts of contemporary geography. So much so geography is understood as the study of the relationship between space and society. The term space refers to the earth's surface or areas (Hartshorne, 1958: 97) and has two connotations - the extent and separation of "things" and differential environmental quality or content (Morrill, Richard, 1985: 63). Society is a sociological concept referring to a network of social relationships. Individuals, groups and institutions constitute the society. There are laws, customs, traditions, and circumstances which govern the pattern of interaction between them. Thus, there are two ways of understanding the nature of the relationship between society and space - spatial organization of society and social organization of space.

Spatial organization of Society

The term spatial organization of society refers to how the society is organized on a geographical space, i.e., the spatial arrangements of people, their activities and institutions, and the linkages between places due to the flow of goods, people, information, etc. More specifically, it deals with the following questions :-

- Where are human activities located on the earth's surface?
- How are human activities distributed on the earth's surface?
- What are the patterns of movement/linkages between places?
- What are the patterns of change in the location and distribution of things on the earth's surface and movement patterns between places?

The geographical space is there, and it is an active force that influences social structures and processes. Space is the key to understand society. It is a spatial determinists or spatialists' view, which means seeing social structures and processes as products of spatial arrangement or material structures (Tonboe, Jens Chr., 1986: 14). Richard Morrill (1985: 263) feels that space matters because of the following four basic facts -

- Space exists.
- Physical and social processes require space to operate.
- With respect to carrying on human activities, space (as the environmental quality/content) varies in content and utility.
- Phenomena, including people, are place-bound - they cannot be everywhere at once.

Morrill further points out that space, both in the form of extent and separation of "things" and in the form of differential environmental quality or content, conditions physical and social activities and processes. More specifically, space influences the operation of social processes as 'social relations are constituted through space, and constrained and mediated by space (Wolch, J. and Dear, M.,1989: 9). The constitutive, constraining, and meditative roles of space are:

- *Social relations are constituted through space*: It means that social relations are not only constituted differently in different types of settlements (urban and rural) but are also constituted in different ways in different parts of a settlement (in private and public places, within households, in streets, university campuses, markets, factories, fields, etc.).
- *Social relations are constrained by space*: It implies the inertia imposed by an absolute built environment or the degree to which the physical environment facilitates or hinders human activity - as in the case of natural hazards.
- *Social relations are mediated by space*: It implies the general action of the "friction of distance" facilitating the development of a wide variety of social practices, including the pattern of everyday life or the development of homogeneous belief systems in geographically isolated communities).

The social organization of space

The term social organization of space accords importance to society. According to this view, the society organizes/creates/produces space for itself by physically modifying it or assigning functions to space, or giving it some meaning and significance. It is also pertinent to note that society organizes space in various ways to achieve some predefined objectives. Society is neither neutral nor apolitical. The way society organizes the space depends on the existing pattern of power relations in the society. The powerful social groups organize space in their favour. For example, those influential in the society may locate certain human activities on the space to influence/define/control patterns of social interaction?

Similarly, the society's elite may define and give meaning to space to include/exclude specific people. Therefore, space (location, distribution, movement, or spatial change) needs to be understood as a product of existing social structure and power relations as per the reflectionists' view. It means seeing the (space) spatial arrangement as a reflection (outcome) of some social structures and processes (Tonboe, Jens Chr., 1986: 14).

At this point, it becomes appropriate to quote what the British Prime Minister Winston Churchill (1943) said on the importance of the shape of the Commons Chamber in his address to the House of Commons: 'we shape our buildings, and afterward, our buildings shape us' (Source: UK Parliament). Churchill is taking a spatial determinist view here, which implies that "We shape (spatially organize) our places (houses, buildings, villages, and cities) with an intention that our place should shape us." How are we shaping (spatially organizing) our cities? We are shaping, for example, our cities by:

- widening roads
- constructing flyovers, sub-ways, over-bridges,
- clearing slums
- beautifying landscape, and
- developing dedicated roads (e.g.,rickshaw-free roads).

What are the implications of the changes? These changes are deliberately made to shape patterns of interpersonal and inter-group relations within the city.

Dear and Wolch (1989: 9) have joined the two views, i.e. spatial separatist and reflectionist, to understand the working of the contemporary world. They believe that 'the pattern of human territoriality (human use of space) results from the interaction between the economic/political/social spheres (of people) and the constitutive/ constraining/meditative role of space' (Dear and Wolch, 1989: 9). One can infer from the entire debate that to understand the pattern of human territoriality, we need to take both the spatial determinist as well as reflectionist viewpoints into consideration.

Society-space-map interface

Society, as a network of social relations, is an active force. It constantly modifies and produces space. In the process, society leaves its marks on space. As a map is a graphical presentation of space, it can be considered a mirror of society. However, as discussed above, the relation between the map and the world is a complex one. The existing power relations in society mediate mapping to such an extent that maps do not always correctly represent the spatial organization of society. The state and the elite use them to express their views. Conversely, mapping can also be a communicative tool to counter the dominant narratives on place created by conventional maps. Thus, they can create not only alternative worlds but also provide alternative viewpoints of the world.

Visualizing the world as it suits political agenda

Before World War II, cartography was used for political propaganda in Europe, particularly by the Nazi Germany. The Nazi used the map to spread half-truths and lies about Germany and other countries of Europe to the world. For example, the Nazi's maps showed how their enemies surrounded Germany. These maps were also used to mobilize public opinion in favour of the rulers. In addition, they promoted the doctrines of racism, expansionism, and authoritarianism (for details, refer to Mason, Betsy, 2016). These maps, called variously such as persuasive, suggestive, or rhetorical or didactic cartography, and sometimes propaganda mapping, provide an edge, great or small, in communicating the intended message: religious or political, military or commercial, moral or social (for details, refer to Mode, P. J., 2017).

Visualizing the world as it appears

Modern cartography aims to visualize the world objectively. In the process of organizing space, society leaves its marks/intention on it, and hence, the former is a mirror of the latter. Space reflects the social structure and pattern of relations within the society. For example, the spatial elements (e.g., farm size, farm structures, dwellings types and spacing between houses, the density of houses, roads, agricultural technologies, etc.) constitute the landscapes where subsistence agriculture is practised and how it is fundamentally different fromthose of commercial agriculture. Thus, geographers, in mapping space, actually map society. Therefore, geographers' study of the organization of a territory is not an end in itself. Instead, it is a means to understand the nature of social structure and power distribution in society?

Modern cartography developed after World War II to objectively portray the world. Arthur Howard Robinson (1915-2004), the American geographer and cartographer, was one of the most significant scholars of modern cartography. He worked in the Map Division of the Office of Strategic Services (OSS) during and after World War II. His job was to develop map-making capabilities and map intelligence for wartime strategic analysis and planning (Joel, Morrison, 2008). Subsequently, he was appointed as the Chief US Map Officer and awarded the Legion of Merit from the US Army in 1946. In 1947, he joined as a Professor in the Geography Department at the University of Wisconsin-Madison). His famous book, Elements of Cartography, was published in 1953, is in its 16th edition. Modern cartography aims to make maps that are reliable, clear, efficient, and effective communication devices. They are based on empirical data and scientific grounds. They strictly follow the rules of cartography. These maps intend to represent the earth accurately; therefore, are considered objective and ideologically neutral, and apolitical.

However, modern mapping failed to capture the rules of the society as it remained elusive to the role of power in creating unequal worlds. In this direction, the works of J.B. Harley (1989) are ground-breaking. He defined maps as social constructions as maps contain knowledge that is socially constructed. It is also essential to note that maps are political as they represent and reinforce power relations. Therefore, he considers map-making as an ideological exercise as maps represent existing power relations in society.

However, maps are not only representations of existing power relations. They are deliberately constructed within the society to perpetuate and reinforce unequal power relations. The mapping activities occur within the power dynamics of society and are hence influenced by it. Thus, as already stated, maps are constructed to represent the world as wished by the elite.

Visualizing alternative worlds

Modern cartography is criticized for merely mapping the surface reality - the visible one. They represent apparent reality but not fact. It uses data that is often collected uncritically to map the places and the world as a whole. However, truths are often hidden or made to hide. Space tends to conceal social reality or is used to hide facts. At times, space is also intentionally used by powerful interests to hide the truth and suppress it. Given that, the task of a geographer is to uncover the hidden social reality by mapping space.

During the 1980s, criticism of scientific cartography emerged, and since the 1990s, many works on critical cartography have been produced.

Critical cartographers accept maps, but the authority claimed by normative maps to portray reality without biases, i.e., maps as truth. Critical cartography is a critique of modern (also called conventional) cartography. It questions the assumptions of professional cartography- maps are scientific documents, and they are neutral representations of the world. Critical cartography challenges academic cartography by linking geographic knowledge with power and thus is political (Crampton Jeremy W. and Krygier, John, 2006: 11). At the same time, critical cartography aims to reveal 'the hidden agendas of cartography' as veils as well as tools of socio-spatial power. In this sense, critical cartography emerges as an alternative way of visualizing the world. It is a way of visualizing alternative worlds.

Visualizing the world alternatively

Critical Cartography opens up possibilities for new forms of knowledge production. There are multiple traditions. According to Suely Rolnik (1998), critical cartography presents the choice of new worlds, new societies. The following are some examples of the recent cartographic traditions which talk of new worlds -

- Counter-mapping refers to any map-making processes that challenge the formal maps of the state. It includes indigenous cartography or ethno-cartography, art mapping, and cognitive mapping. Nancy Peluso (1995) used this term to describe mapping practices by indigenous people in Kalimantan, Indonesia, as they made maps to contest Indonesian state land-use plans.

- Indigenous mapping, sometimes referred to as Ethno-cartography,is non-western mapping practice created from the perspective of indigenous communities and running challenging colonial assertions over land claims.

- Parish mapping refers to the large-scale and local mapping of areas, usually done for activist purposes or collecting and mapping local knowledge.

- Participatory mapping aims to understand the connections between people, places, and organizations over space based on indigenous conceptions of territories and world views.

- Everyday mapping aims at mapping everyday experiences, such as fear, insecurities, comfort, ecstasy, etc., of various social groups, i.e., male, female, children, and aged.

- Autonomous cartography refers to the cartographic practice that is free from the institutional framework.
- Alternative cartographyis an alternative method of visualizing the world as well as a way of visualizing alternative worlds
- Cognitive mapping or mental mapping refers to how people comprehend, learn, remember, record, and articulate their experiences in the physical environment.
- Eco-mapping is a mapping for ecological purposes, usually at the grassroots by environmental activists. It is similar to green mapping.

Visualizing the world for changing it

Denis Wood and John Fels (2008) claim that maps are constructions that produce the world. According to them, maps make reality as much as they represent it. Maps are not passive but active and exercise power. They are constructed to produce the world and can do things. Moreover, maps open up our thinking spaces, and by doing so,they can encourage us to ask for a new world. In this way, maps can also be powerful tools for promoting social change. Mapping can also be viewed as an alternative way of visualizing a better world.

Conclusion

Mapping is an age-old practice. Maps have alwaysbeen of help to humans in making sense of the world since the dawn of civilization. The modern mapping practices are based on the premise of scientifically and objectively representing the earth. However, in recent decades, critical cartography has emerged as a critique of modern mapping practices. It questions the very assumptions of scientific maps and the science of mappings. Critical cartography claims that modern mapping fails to reveal the social reality hidden in space. It calls for alternative ways of visualizing the world as well as for the visualization of alternative worlds. Further, maps are active and constantly keep on shaping our thinking process. Given this, alternative maps are needed that can act as political interventions in the world. These maps will depict the unjust worlds, which otherwise remain unrepresented. At the same time, they will visualize a spatially organized world on the principle of justice.

Acknowledgments: The author places on record his sincere thanks to Dr. Taruna Bansal, Associate Professor, Department of Geography, Jamia Millia Islamia, New Delhi, for her valuable suggestions on this paper.

References :

Betsy, Mason (2016). All over the map: these colourful propaganda maps fuelled 20th-Century wars, The National Geographic. October 5, 2016. At https://www.nationalgeographic.com/culture/article/propaganda-war-maps-gallery; Accessed on July 27, 2021.

Crampton Jeremy W. and Krygier, John, (2006). An Introduction to Critical Cartography. ACME: An International E-Journal for Critical Geographies, 4 (1): 11-33.At https://www.acme-journal.org/index.php/acme/article/view/723/585; Accessed on July 23, 2021.

Churchill and the Commons Chamber - UK Parliament (1943). At https://www.parliament.uk/about/living-heritage/building/palace/architecture/palacestructure/churchill/;Accessed on July 22, 2021.

Dear, Michel and Wolch, Jennifer (1989). How territory shapes human life. In Wolch, Jennifer and Dear, Michel (eds.). The Power of Geography. Boston, Unwin Hyman Inc., pp. 3-18.

Harley, John B. (1989). Deconstructing the map. Cartographica, 26 (2): pp. 1-20.

Hartshorne, Richard (1958). The Concept of Geography as a Science of Space, from Kant and Humboldt to Hettner. Annals of the Association of American Geographers, 48 (2): pp. 97-108.

MAP by Wislawa Szymborska, translated, from Polish, by Clare Cavanagh. The New Yorker (April 14, 2014); At https://www.newyorker.com/magazine/2014/04/14/map; Accessed on July 15, 2021.

Mode, P. J. (2017). " Not maps at all" - What is pursuaive cartography? And why doe it matter? The Portolan, 100: 8-26; At https://persuasivemaps.library.cornell.edu/sites/default/files/PJM%20Persuasive%20Cartography%20Portolan%20%23100%20Winter% 202017.pdf; Accessed on July 27, 2021.

Morrill, Richard (1985). Some important Geographical Questions. The Professional Geographers 73 (3): 263-270

Morrison, Joel (2008). Arthur Howard Robinson, 1915-2004. Annals of the Association of American Geographers, 98 (1): pp. 232-238.

National Geographic Society (2021). Prime Meridian, at https://www.nationalgeographic.org/encyclopedia/prime-meridian/; Accessed on July 27, 2021.

Peluso, Nancy (1995). Whose woods are these? Counter-mapping forest territories in Kalimantan, Indonesia. Antipode, 27 (4): pp. 383-406.

Robinson, Arthur H., Morrison, Joel L., Muehrcke, Phillip C., Kimerling A. Jon and Guptill, Stephen C. (2009). Elements of Cartography, New Delhi, Wiley India Pvt. Ltd.

Rolnik, Suely (1998). Sentimental Cartography. Translated from the Portuguese by Adrano Pedrosa and Veronica Cordeiro; At http://www1.uol.com.br/bienal/24bienal/rot/txt_ing_ensroln.htm. Accessed on July 20, 2021.

Tonboe, Jens Chr. (1986). On the Political Importance of Space the Socio-Spatial Relations of Trade Unions, Gender, and the Decentralized Danish Welfare State. Acta Sociologica, 29 (1): pp. 13-30.

Treib, Marc (1980). Mapping Experience, Design Quarterly. 115: pp. 1+3-32. At https://www.jstor.org/stable/pdf/4091019.pdf; Accessed on July 21, 2021.

Wood, Denis and Fels, John (2008). Commentary: The Nature of Maps: Cartographic Constructions of the Natural World. Cartographica, 43 (3): pp. 189-202. At http://www.deniswood.net/content/papers/Wood%20and%20Fels-Natures.pdf; Accessed on July 19, 2021.

11

Unorganised Sector Workers during Pandemic and Role of Government Entitlements

Ashish Ranjan
ashish.ranjhanjha@gmail.com

National Level Lockdown imposed in India on 24th March, 2020 in the wake of Covid-19 pandemic was harsh, sudden and unplanned. This resulted into a big crisis for the unorganized workforce in India. Unorganized sector workers, also known as informal sector workers have no work protection (can be fired easily, no work protection during health emergencies etc.) or social security (pension, gratuity) protection. The Economic Survey of 2018-19, released on 4th July 2019, said almost 93% of the total workforce is "informal". It is estimated that as all major economic activities came to a halt, unorganized sector workers were thrown out of work and stripped off their income. Millions of migrant workers were left stranded in their place of work without any source of income. Farmers were unable to sell their produce. All of this resulted into a big humanitarian crisis. Thousands of migrant workers started returning to their native places from their places of work as they did not have sufficient food to consume. Economic vulnerabilities and food insecurity increased substantially. A number of survey reports have shown that livelihoods were devastated at unprecedented levels. A survey of 5000 self-employed persons conducted between 13th April and 23rd May, 2020 by Azim Premzi University (APU) in collaboration with civil society organizations found that two third of the respondents had lost work. Few informal sector workers who managed to retain their employment during the lockdown saw their earnings drop by more than half. Almost 8 out of every 10 persons were eating less food than before and more than 6 out of 10 persons in urban areas did not have money

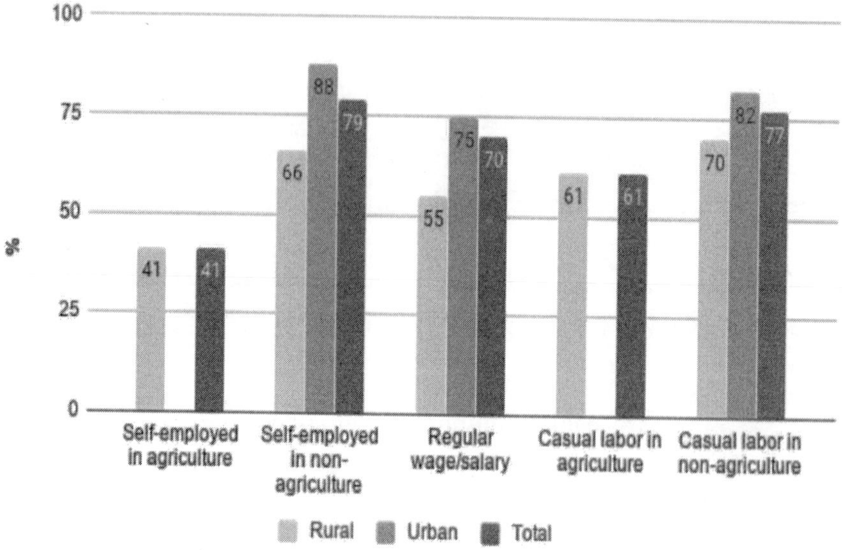

Figure 1: Loss of Employment by region and activity status

Source : https://cse.azimpremjiuniversity.edu.in/wp-content/uploads/2020/10/Kesar_et_al_Pandemic_Informality_Vulnerability_June_2020.pdf

for weeks' worth of essentials. The survey also found out that the impact of job losses and food insecurity was higher for certain groups of people such as Muslims, Dalits, Women, and those with lower level of education.

The survey also revealed that the relief measures announced by the

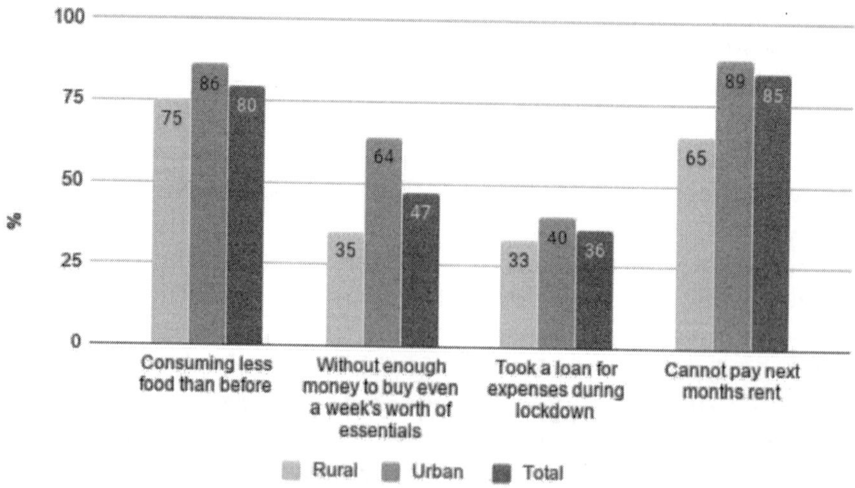

Figure 2: Impact of lockdown on households (%)

Source: https://cse.azimpremjiuniversity.edu.in/wp-content/uploads/2020/10/Kesar_et_al_Pandemic_Informality_Vulnerability_June_2020.pdf

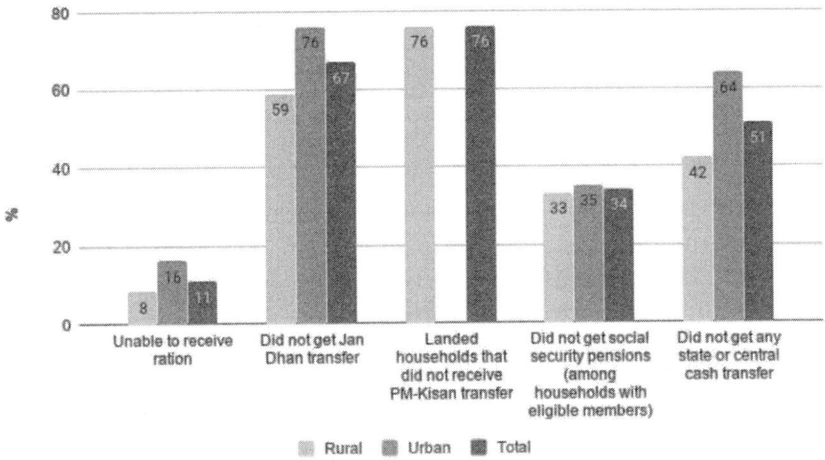

Figure 3:

Source: https://cse.azimpremjiuniversity.edu.in/wp-content/uploads/2020/10/Kesar_et_al_Pandemic_Informality_Vulnerability_June_2020.pdf

government were not in proportion to the severity of the problem. Bulk of the stimulus package was directed at improving the liquidity rather than direct spending. Half of the respondents did not receive any cash transfer.

Even when compared with the well-off countries like the USA, which spent about 6% of its GDP on direct benefits, India lagged behind with only about 1.5% of GDP spent on direct benefits.

Right after the lockdown noted developmental economist Jean Dreze suggested making good use of existing social security schemes to support poor people. He listed Pensions, midday meals, MGNREGA in the article that he wrote for The Hindu[2]

Indian government's response, as noted earlier, was not enough but Govt. of India did make use of the existing social security schemes to address the situation. Two of the central government's schemes that were most talked about and were given the most attention in terms of government funding, were the Public Distribution System (PDS) and Mahatma Gandhi National Rural Employment Guarantee Act (MGNREGA). Both these schemes are covered by "rights based legislation" which makes the entitlements enforceable by law.

National Food security act was enacted in 2013 and provides for a public distribution system. Public Distribution System covers around 66% of the population - 75% in rural areas and 50 % in urban areas.

About 80 crore Indians are covered under this scheme. In Bihar, about 84% of the population is supposed to be covered under this scheme, making it almost universal. Each beneficiary gets 5 Kg of grain at a subsided rate of Rs 2/3 per kilogram.

Government of India doubled the allotment and decided to provide 5 kgs of grain per person free of cost for six months, over and above what was already being given under the provisions of the National Food Security Act (NFSA). One kilogram of pulse was also given free for each household along with the grain. The logistical nightmare of quickly reaching out to people with food grain did not seem to be a challenge as an elaborate mechanism of purchase, storage and distribution was in place for a number of years. This was a welcome move and brought much needed relief to people. This release of food stock was also needed to empty the FCI Godowns for procurement of rabi crops. Still a large number of people who are not covered by the PDS due to various reasons were left out. In Bihar for example only 70% of the population is covered as opposed to the mandated 84% under the NFSA. The reason for this is that numbers have been calculated based on the 2011 census (10.38 crores) and there was some shortfall in allocation by the centre. If one takes into account the projected population of Bihar in Dec 2020 (about

Population of Bihar, 2011 (lakh): (Source: Population census 2011)	**1,041**
Population of Bihar, 2019 (lakh): (Source: https://www.census2011.co.in/census/state/bihar.html)	**1,257**
Mandatory PDS coverage under NFSA (%) (rural 85%, urban 74%)	**84**
Mandatory coverage in absolute terms (lakh persons)	
Using 2011 population	872
Using estimated 2019 population	1,053
Population actually covered by PDS (https://dfpd.gov.in/coverageandallocationundernfsa.htm)	**857**
Amount of extra foodgrain required to cover the following groups @ 5 kg/person/month for 12 months (lakh tonne)	
Population left out due to population under-count (1,053 lakh - 857 lakh = 196 lakh)	11.7
Entire rural population and urban slums* (1,218 lakh - 857 lakh = 361 lakh)	21.6

Figure 4: Estimates of Extra Food Grain in Bihar

*The 1,218 lakh figure is based on the following assumption: (1) total population in 2019 = 1,257 lakh (see above); (2) rural population = 12% of total; (3) urban slums account for 74% of urban population. It is also being assuming that current PDS coverage in urban areas is confined to urban slums.

12.5 crores), then it comes out that roughly 70% of the population gets benefits under NFSA (Figure 4). Undercount because of increase of population from 2011 itself comes to about 2 crores. It can be calculated easily that in order to cover all households living in rural areas and all urban poor in Bihar (people living in slums), it would require only 2 million tonnes of grains.

The Central Government could have easily released this stock because it had in its stocks 80 million tonnes of grains in March 2020. Unfortunately, this could not be done leaving even those entitled without any food support from the government. It was also suggested that government open food kitchens in urban centers where people could get hot meals and avoided hunger. School infrastructure was already in place to provide cooked meals to hungry people in villages and that could have been utilized as well. Unfortunately these measures could not be taken up at a large scale. Though, Government of Bihar did run few relief camps in each districts where meals were served. What is striking is while FCI godowns were overflowing with foodgrain stocks, crores of people who were entitled could not be supported.

Halt in several economic activities left people without work. MGNREGA provided some relief. MGNREGA saw unprecedented jump in the demand levels compared to previous years. There was 47% increase in the number of households that demanded work in MGNREGA in Bihar between April and June (Table 1). 48% of the MGNREGA budget was spent in the first four months of the 2020 financial year, i.e. April to July. There was a 43% increase in the persondays generated till November 2020 when compared to the same period of the previous year 4. The government of India allocated additional Rs 40,000 crore (taking the total budgetary allocation to around 1 lakh crore in 20-21) in one of its stilumus package and promised to allocate more if needed. Still there was an unmet demand of 13 % at the national level and 17% at the Bihar level 6. Till February, 2021 around 48 lakh households have worked in MGNREGA in Bihar and had received on an average 42 days of work. In 2019-2020, 34 lakh households had got work in Bihar. This means that in the pandemic year, 14 lakhs more households have got MGNREGA work - a 41% increase (Data source: MNREGA MIS at nrega.nic.in). Rise in the work demand corroborates the loss of work among the workforce. The work demand was not only from the migrants who had lost work but also from all quarters as the incomes of people had fallen drastically during the lockdown. Even after the increase in Budget and subsequent increase in persondays generated it was widely reported and felt that only a fraction of population could avail MGNREGA work[3].

Table 1
Increase in registered household demand for work under MGNREGA in Bihar

	No. of households demanded work 2019-20 (April-June)	No. of household demanded work 2020-21 (April-June)	Decrease in per cent
No. of households demanded work	5521959	3763143	**46.7%**

Data Source: MNREGA Public Data Portal https://nregarep2.nic.in/netnrega/dynamic2/dynamicreport_new4.aspx

There is no denying the fact the income generated from MGNREGA, helped poor households avoid extreme hunger. Thus, it can be deduced that MGNERGA should be expanded further and effort should be made to increase the average number of work days. Bihar has a huge population of agricultural workers and hence the potential for expanding MGNREGA in Bihar is huge. One big concern for the workers is that the MGNRGEA wage is far below the statutory minimum wage in Bihar. Since the scheme is centrally funded it is expected from the central government to honour the minimum wage in Bihar. Bihar government can also open its coffers and offer the statutory minimum wage in NREGA. This will further strengthen the programme and also encourage participation.

Conclusion:

Informal sector workers were hit most by the economic impact of the pandemic followed by the lockdown. The economic package provided by the Government of India during the time of pandemic falls short in addressing the kind of pandemic induced economic impact on vulnerable sections of the society. A large percentage of population has not received any cash transfer or other benefits from the government. Nonetheless, both public distribution system and employment guarantee programme (MGNREGA) came to the rescue when people needed it most. Fortunately for the country, both these entitlements, namely NFSA and NREGS were already in place and were a mature programme when the lockdown was announced. There has been much criticism of food subsidy and NREGS by a section of economists who think that subsidy could be used in more productive ways. However, we have seen that these two programmes were critical in addressing food insecurity situation in the most difficult times. What is needed is to further strengthen the two programmes by including those who have been left out of the counting in PDS, by providing additional grains in the time of crisis and by providing adequate resources to MGNREGA so that everyone can be provided work with dignity.

References:

Dreze Jean, "Excess Stocks of the Food Corporation of India must be released to poort", April 9, 2020, https://indianexpress.com/article/opinion/columns/coronavirus-lockdown-food-for-poor-migrants-mass-exodus-jean-dreze-6353790/

Dreze Jean, "Perils of an all out lockdown", March 23 2020, https://www.thehindu.com/opinion/lead/the-perils-of-an-all-out-lockdown/article31136890.ece

Kapoor Mahima, "Only 20% Surveyed Found Jobs under MGNREGA During Lockdown", August 10, 2020

https://www.bloombergquint.com/economy-finance/only-20-surveyed-found-jobs-under-mgnrega-during-lockdown-gaon-connection

Pandharipande Neerad, "MNREGA during COVID-19 pandemic: More households got work in 2020, but fund crunch a major concern, finds study", December 21, 2020, https://www.firstpost.com/india/mnrega-during-covid-19-pandemic-more-households-got-work-in-2020-but-fund-crunch-a-major-concern-finds-study-9162261.html

BhaduriAmita, "NREGA tracker looks at the schemes' performance", 26 july, 2020 https://www.indiawaterportal.org/articles/nrega-tracker-looks-schemes-performance

"PAEG Report on status of MGNREGA", December 2020, https://www.indiaspend.com/uploads/2020/12/23/file_upload-364421.pdf

12

Knowledge Economy, Labor Market and Spatial Science: Challenges and Opportunities for Bihar

Rajiv R. Thakur (Ph.D.)
rajivthakur@missouristate.edu

Introduction

If we look around the icons of knowledge economy is all around us in myriad shapes and forms. Be it our smartphones, laptops, iPad, or how we purchase goods on amazon, buy our flight tickets and pay our utility bills. Understanding knowledge economy, labor market and how these interfaces with spatial science and its tools and techniques, particularly in the context of Bihar is significant as the forces of globalization and its influence is changing the fortunes of the local and regional economies. More so, the local and regional economies and associated labor market, as well as the collective dependence upon social and physical infrastructures (transportation networks, utility grids, as well as capacity of local governments) are all fast changing. This requires an appreciation of the conceptual foundation of how resources are drawn from the natural environment, incorporated into economic processes, and are assigned an economic value. In doing so, it is important that we understand the fundamentals of market system, and how it inevitably creates an uneven geographic landscape. In the context of Bihar, the author unpacks the forces of globalization and its influence on its landscape of economic geography, acknowledging the influence of global economic restructuring as has unfurled in the last few decades.

[1]In response to an invitation from the Center for Geographical Studies, Aryabhatta Knowledge University, Patna the author had the opportunity to share his thoughts on the subject through a lecture/talk on July 24, 2021. Presented here is an expanded version of the lecture.

Resource Based and Knowledge Based Economy

Like many other places in the world and India, Bihar's economy is characterized by extraction of resources or very simply transformation of nature i.e., land into something useful. Embedded in this process are social institutions and structures to facilitate the same. Thus, agriculture is the mainstay and primary source of livelihoods in large parts of rural Bihar. The ways in which nature is harnessed and incorporated for economic purposes is obvious: farming for subsistence or exchange. At its very best the engagement through agriculture and related activities are mediated by exchange relationships, for example in a variety of retail, and related utility services. As such, the dynamics of resource-based economy is dependent on low levels of technological and scientific knowledge, creating low value additions. In Bihar, where approximately 25% of its workforce is dependent on production of plant crop, the geography of its production, and the social system that supports it, is shaped not just by its ecosystem characteristics but also by the nature of the cultivation process itself. Finally, in the context of Bihar if resource-based extraction must continue then it also calls for a certain level of investment, infrastructure and logistical support which means that the role of government is central for the continued success of resource-based economies. As we examine all these features of commodity creation and transformation, it becomes obviously clear that it will lead to a very distinctive pattern of uneven development characterized by changing establishment of ownership in the form of communal access, state ownership and exploitation, state ownership and private exploitation, private ownership and private exploitation.

Contrary to the above arguments associated with a resource-based economy is the existence and evolution of knowledge-based economy which presupposes the existence of western style industrialization and the process of deindustrialization as was experienced in the late 1970s and 1980s because of economic restructuring. A typical knowledge regime or economy is one which is characterized by knowledge-intensive manufacturing or services. In such an economic regime the skill and expertise of workers and the innovation that it facilitates lie at the heart of the success of the firms and hence the region itself. Conceptually, Peter Drucker (1969) is the source of the idea of 'knowledge worker'. While in the western world this notion of a knowledge economy has been the backbone of the process of deindustrialization and the shift towards skill development and training of workers leading to more competitiveness because of innovations. More advanced conceptualizations of knowledge economy which include notions of clusters, learning regions, industrial districts and innovative milieux. Whatever be the case, competitiveness and innovation remain central to knowledge economy and to that effect may even drive how capitalism and market economy evolves and thrives (Van Geenhuizen and Nijkamp 2012; Audretsch and Walshok 2013).

However, many underdeveloped regions worldwide and/ or peripheral regions like Bihar escaped both industrialization and deindustrialization in the western sense and continued to grow and survive. Many policy and economic commentators heralded regions like Bihar and its unparalleled growth as signaling the emergence of a 'new economy', in which economic growth and success had less to do with the production of tangible commodities and goods and much more to do with the production and deployment of knowledge. Indeed, the term knowledge economy and 'the new economy' quickly became synonymous. To that end, the above narrative is an attempt to clarify both the conceptual notion of 'knowledge economy' and display that knowledge economy interfaces with the global and the local and is about spatial interconnections, interdependence, and interactions.

Knowledge Economy and Labor Market

Unlike a typical resource-based economy, by definition and its inherent logic, knowledge economy and the associated labor market is more like an upside-down pyramid where the workforce participation in primary sector is the lowest and progressively increases as one moves up the economic sectors i.e., from primary (farming, fishing, mining etc.) to quaternary (business service provider or those providing services to the service sector). A defining characteristic of this upside-down pyramid is that the value addition in monetary terms increases as one moves from primary to quaternary sector. Also, unlike the popular misconception, knowledge economy does not necessarily simply involve information technology at its best, but constantly evolving skills which drive innovation and further pushes competitiveness in an economic system (Audretsch and Walshok 2013). Interestingly, while the skill and labor power of a worker and/or capitalist in resource-based economy is a zero-sum game given the inherent nature of how capitalism and commodification work i.e., land or such other resources can only belong to one person at a time, in a knowledge landscape the skills and entrepreneurial capacity of workers are in a positive-sum game i.e., the same skill sets can be had by more than one person (Figure-1). All this speaks to the fact that knowledge landscape is a game changer. In a sense, traditional notions of scale and hierarchy in economic landscape

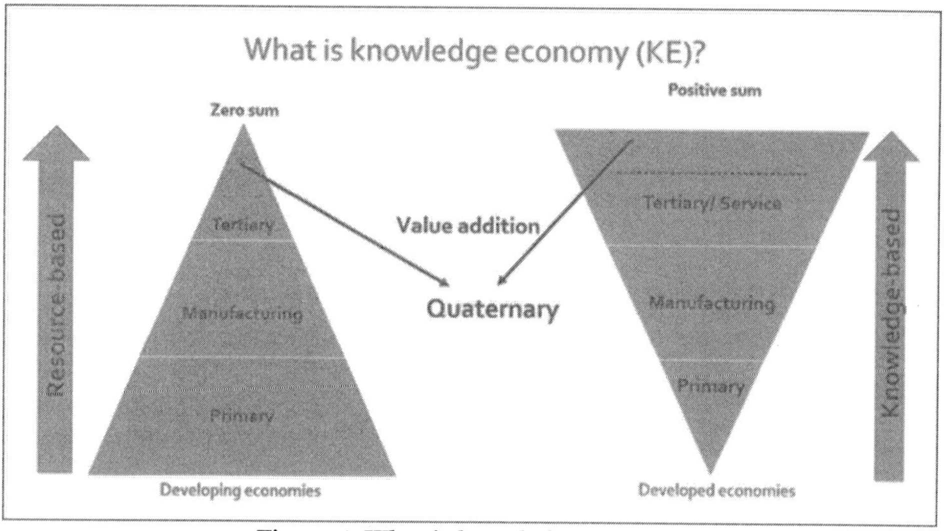

Figure-1: What is knowledge economy

will see itself changing dramatically. This may change our traditional understanding of commodification in a market society.

Spatial Attributes of Knowledge Economy

As we dive to understand economic thinking and development within knowledge economy, it becomes amply clear that features such as entrepreneurialism, universalization of technology and constant innovation are all contributing to the shrinking of space, rewriting how we view commodification and market society. Some changing spatial attributes include -

- Knowledge economy (KE) promises reversing the traditional notions of doing business i.e., moving us from concentration to decentralization meaning we need to ask questions such as how and where do we shop?
- KE promises to bring closer both production and imagination. While proximity still matters, technology is certainly leading to shrinking of space.
- KE changes the moral culture of production
- KE has led to secular stagnation. Broadly speaking, capital controls creation of deepening gap.
- In a sense KE has brought a dent in legal and institutional structures causing reinvention of the market economy

While the landscape of KE gives us the perception of localization, freedom, shrinking of space, among others, from a social and intellectual context we are falling under more command and control than ever before.

Knowledge Economy and Implications for Spatial Science/ Geography

Granted that KE may create a landscape where certain technologies can be used to transcend time and space and generate both new products and new production processes, however, given the inherent nature of capitalism which is that it creates development in some places and underdevelopment in others, spatial decision making will always have a place even as the geography of technology production is itself highly uneven. In other words, whether it is schools, hospitals or banks, spatial data which involves people and places will always go through the cycle of a new crisis or problem, assessment, and the need for collecting new information, generating alternatives, policy action, and evaluation of outcomes.

Spatial science has emerged as a popular interdisciplinary perspective and is situated between the human and physical realms to provide support to spatial decision making. Thanks to the suite of geospatial technologies available to us in the form of global positioning systems, geographic information systems, and remote sensing which play across the sciences. The notion of spatial science - as a way of knowing the world - continues to evolve. Imagine a wide variety of functions such as understanding facility management, geoprocessing modelling, topographic analysis, site selection, watershed analysis, land management, network analysis, incident mapping, logistic routing among many others is some of the things that geospatial tools and techniques can deliver to facilitate quality well being of people in distant places. This growing awareness of emerging geospatial tools and techniques embedded in intellectual and conceptual vigor, and enhanced analytic abilities are critical to the competitive advantage of individuals, institutions, and regions, as well as, have a role in the preparation of spatial decision making (Gatrell et al 2020).

The most significant element of KE is human capital. The quality of the workforce in the labor market is in direct correlation with its education and skills. For the workforce to remain cutting edge, in response to shifting forms of global integration, continued dynamism in the development of new technology, and the need to address environmental challenges, local and regional economies must be prepared to provide capacity building responding to constant evolution in technology, flexibility of the market through partnerships between universities, trade institutions, and government among others (Thakur 2009).

Challenges and Opportunities for Bihar

What does this mean for Bihar? The landscape of KE requires not just understanding the macroeconomic processes and patterns but also appreciating what it takes for local and regional economies to remain critical in shaping the conditions, prospects and well being of the people of Bihar. Stakeholders will have to appreciate understanding the significance of experiential learning and the need to articulate backward and forward linkages of institutions and processes to facilitate the same.

A few challenges or hurdles for Bihar in facilitating the landscape of KE includes its population dynamics, the debilitating impact of caste as an institution, constant resurfacing, and reproduction of semi-feudal social structure. Not with standing these challenges, Bihar has opportunities in the form of resilient human capital while drawing upon its ancient legacy of knowledge and learning, the opportunities of understanding and articulating an evolving tax incentive structure to attract capital and creation of special economic zones ensuring political stability (Bandyopadhyay 2009; Bhattacharjee 2013; Kumar 2012; Mahajan 2015).

Conclusion

The dynamics of knowledge economy, labor market and spatial science are indeed interconnected. The above discussion displays those global influences have led to change in the economic landscape in far and remote places including Bihar. Appreciating and articulating the spatial dynamics of knowledge economy is important to understand the scale and scope of the social and economic changes that require urgent action. Smooth evolution for Bihar towards a landscape of knowledge economy also involves appreciating institutions and people who would facilitate through growth of human capital. This involves appreciation of the role of spatial science and its attendant tools and techniques which play a significant role in the development of knowledge-based economy and its relationship with environment.

References :

Audretsch DB and Walshok ML (2013) Creating Competitiveness: Entrepreneurship and Innovation Policies for Growth, Cheltenham, UK: Edward Elgar

Bandyopadhyay D (2009) Lost Opportunity in Bihar, Economic and Political Weekly, XLIV (47): 12-14

Bhattacharjee G (2013) Special Category Status: Will it Actually Benefit Bihar? Economic and Political Weekly, 48(18): 17-18

Gatrell JD, Bierly GD, Jensen RR, Thakur RR (2020) Research Design and Proposal Writing in Spatial Science, Switzerland: Springer Nature.

Kumar A (2012) Fallacy of the State in Bihar, Economic and Political Weekly, 47(44):23-25

Mahajan S (2015) Decay and Darkness in Bihar, Economic and Political Weekly, 50 (2): 78-79

Thakur RR (2009) Science and Technology Indicators in EPSCOR States: A Policy Geography, Unpublished Ph.D. Thesis, Indiana State University, Terre Haute, Indiana.

Van Geenhuizen M and Nijkamp P (2012) Creative Knowledge Cities: Myths, Visions and Realities, Cheltenham, UK: Edward Elgar.

13

Psychology of Climate Change: Changing Human Behaviour, Changing Lifestyles and Cutting Consumption to Curb Carbon Emission

Habibullah Ansari
hbansari@yahoo.com

Abstract

Climate change has become a global problem and the scientists of the world are working to solve this problem. The degradation of the environment has occurred because of the unmindful over exploitation of natural resources by human beings. Unnecessary excessive use of resources whether it is food, clothes, cars, and other home appliances required which are somehow forced by the modern market forces and advancement of technologies. It could be stopped by bringing positive changes in human behaviours, the human mind, and by taking collective responsibility. A group of psychologists has worked towards this goal by conducting various studies and have suggested certain ways to curb it. This paper, based on secondary sources, tried to explain the role of psychology to understand the climate change, its impact on human life, and in what ways it could be mitigated? It tries to suggest changes in human behaviours which could be brought to cut the overconsumption of resources and to develop pro-environmental behaviours.

Keywords: Climate change, psychology, human behaviour, cutting consumption, change in lifestyles.

Introduction

Climate change is related to global warming which is the result of the emission of a large amount of carbon and other gases in the atmosphere. Directly and indirectly, it is related to human behaviour such as over-consumption of natural resources and over-production of the manufactured goods for human consumption even beyond human needs. Sometimes some material is not required for the survival of human needs but the marketing condition push compel humans and forcing them to adopt a non-sustainable and even costly lifestyle. So if the world is willing to stop global warming it has to change the lifestyles of the people and have to adopt a sustainable, traditional way of life-based on minimum consumption and it has to be satisfied with nature for the sake of being alive on earth. The human being has to forgo the consumerist behaviour which is based on so-called modern and advanced technologies. The human being today has become the slave of machines and technology and lost its potentialities. The people, who can afford to purchase, are enchanted with high tech machines, luxurious cars, air-conditioning, various types of advanced electronic gadgets as well as with high and lofty palaces, high rise buildings (towers), etc. without rational thinking whether these luxuries they require or not or just because of the market forces which is compelling them to adopt those luxurious lifestyles. Thus we have to change our lifestyles by changing behaviours and focus on cutting over-consumption to save energy and reduce the emission of carbon and other gases. We have to adopt the traditional ways of life which were based on self-sustenance. Get what the bare minimum a human being requires to live a happy and self-sustaining life. For that, we have to change the people's mindset, their thinking, change their attitudes and perception towards overconsumption, and protecting and preserving natural resources. If it could not be done, then like today'scorona pandemic everybody will be at risk of life due to heavy pollution and environmental degradation andthey will cry for the fresh air-oxygen. Today you are purchasing water tomorrow you have to purchase oxygen and oxygen may not be available for everybody and at everywhere.

With the above assumptions this paper, based on secondaryand tertiary sources, tries to explain the role of psychology and psychologists to change human behaviour in favour of stopping climate change by reducing carbon and other gases as well as protecting the physical environment. Human behaviour is an integral part of not only to causeof global climate change but also responding and adapting to it. Psychological research should inform efforts to address climate change, to avoid misunderstandings about human behaviour and motivations.

Psychology and Climate Change

Till now, psychology has been largely left out of the conversation about climate change, but researchers in the field of environmental psychology are seeking to change the idea. Climate change is not just a political, social and economic issue. It is also a deeply psychological one and now, behavioural scientists are using psychology to better understand the complex relationship between people and nature. An increasing number of psychologists are arguing that in order to tackle the growing threat to our environment, we need to understand people's emotional and cognitive responses to this new reality, which can run the gamut from denial to indifference to outrage to anger to grief (Gregoire Carolyn, 2016). Scientists in the burgeoning field of environmental psychology are working hard to bring psychological insights into discussions about climate change. Research in environmental psychology can focus on a number of different areas of inquiry relevant to climate change, including the psychological and mental health impacts of ecological crises and disconnection from nature; the use of psychology to inform communication about climate change, advocacy and policy; and the psychological roots of climate change denial, apathy and inaction.

The American Psychological Association (APA) has constituted a task force on the interface between psychology and global climate change met in 2008-09 to examine the role of psychology in understanding and addressing global climate change, including efforts to adapt and to mitigate climate change (APA Task Force Report, 2011). The task force's report reviews a wide range of research and practices relevant to climate change, including work in environmental and conservation psychology, studies of human responses to natural and technological disasters, efforts to encourage environmentally responsible behaviour, and research on the psychosocial impacts of climate change.

Some of the points the APA Task Force has found are the perceptions of global warming and climate change risks, including people's tendency to discount the likelihood of future and remote events and the role of culture in how people conceive and respond to risks. Human behavioural contributions to climate change, such as population growth, energy use, and consumption, are psychological as well as contextual drivers of these contributions. Psychosocial and mental health impacts of actual and perceived climate change, including stress, anxiety, apathy, and guilt, and interventions to promote coping, adaptation, and healthy responses to climate change. Social and community impacts of climate change, socioeconomic disparities in climate change impacts, and ethical and social justice implications of climate change. Psychological barriers that limit individual and collective action on climate change. Empirically-based approaches to understanding the nature and determinants of behaviours that affect the environment and the development of interventions to alter such behaviours (APA Task Force Report, 2011). The members of the task force argued that work in the complex arena of climate change cannot be left to one sub-discipline but must draw upon the expertise of researchers and practitioners from multiple areas of psychology.

A group of psychologists had brought a resolution on Affirming Psychologists' Role in Addressing Global Climate Change. They had claimed that there is a greater role of psychologists to address the issues of climate change. There is a near consensus among climate scientists that global climate change is occurring faster than anticipated, and there will be greater global climate change if greenhouse gases are not reduced. Psychologists can provide a behavioural analysis of such contributions and there is a need for inter- and cross-disciplinary research on Global Climate Change that includes the social and behavioural sciences, and psychologists have been and are collaborators and participants in such research (APA Task Force Report, 2009; Fischhoff, B., & Furby, L., 1983; National Research Council, 2010).

The psychologists have affirmed that the impacts of climate change are increasing globally and include the destruction of habitats and subsequent threats to endangered species, acidity of water, disasters (e.g. forest fires), extreme weather (e.g., hurricanes, heat waves), decreasing availability of water, and spreading of diseases, harming plants, wildlife, human physical heath, settlements, and psychological well-being, and are a threat to social, economic, and environmental sustainability (APA task force report on Psychology and Climate Change, 2009).

Psychologists have shown concern about individual and institutional discrimination (e.g., APA Resolution on Poverty and SES, 2000; APA resolution on Prejudice, Stereotypes, and Discrimination, 2006); and climate change has already had a disproportionate impact on the poor, including greater impacts on women and children, on rural regions and their inhabitants, and is anticipated to have greater effects on already disadvantaged populations including but not limited to persons with disabilities (APA task force report on Psychology and Climate Change, 2009; International Disability and Development Consortium, 2008; National Research Council, 2010).

The APA in its mission and vision statements and its ethical code of conduct indicates that psychologists are committed to creating, applying, and communicating our knowledge to improving individual and societal conditions and facilitating the resolution of global challenges. Psychology as a discipline is well-suited to address important behavioural and methodological aspects of understanding human behavioural contribution and responses to global climate change (Clayton & Brook, 2005, Gifford, 2007; Uzzell & Moser, 2009). The APA is committed to education in psychology and the dissemination of sound psychological science both in and out of the classroom. The APA Council approved a research agenda on environmental problems. Therefore, it is resolved that APA reaffirms its recognition of the importance of psychological aspects of human-environment relations and supports psychologists' involvement in scientific research on global climate change and on the role of human behaviour as a significant contributor to these changes.

So it is evident from the above discussions that there is a greater role of psychology and psychologists to make understand the risk of climate change to the people. Many efforts have been undertaken by psychologists in the world to curb the climate change through bringing the change in human paviour applying various psychological techniques, theories and applications. Several new fields in psychology discipline have emerged such as Environmental Psychology, Ecological Psychology, Climate Psychology etc. So the psychology is contributing largely to the climate change issues in the world.

How Human Behaviour Can Stop The Climate Change?

Why it is important to focus on behaviour change? Behaviour change is becoming more central to international debates and may represent a 'holy grail' for policy makers. There is extensive research and practice which shows that understanding social and behaviour aspects can be a low cost and effective strategy to reducing emissions. At the same time, there are many co-benefits of a carbonneutral society including public health, increased energy security, cost savings for households, potential job creation and better social cohesion and quality of life (Moore Jeanne, 2012). However, changing behaviour is neither automatic nor predictable. To be effective, social and behavioural approaches need to have an impact in the long-term (thus changing habits and values) and involve not only individuals but systems and social practices across all levels of society. Ultimately, the transition towards a carbonneutral society is a societal project-people have to be engaged, informed, be willing to participate and change their behaviour for climate-change mitigation to take place.

Emphasis should be put on placing particular behaviours within their social, cultural and technological contexts. There is a shift away from a purely economic model of consumer choice, to one which recognizes that long-lasting behaviour change needs to occur across all levels, involving institutional, technological change as well as a shift in social practices and norms. Widening and deepening is required in understanding behaviour change-widening in the sense of moving beyond the individual to shared practices, habits and routines- and deepening, in that it is necessary to get under the skin of attitudes, to deeper motivations and values, for long-lasting change (Moore Jeanne, 2012).

The public in general is not very much aware about the climate change. The issues related to climate change have the discussion point among the intellectuals and some activist groups. In the USA a poll survey was carried out among the common people to rate the climate change problem in 2014 by the Pew Research Centre (Lombrozo Tania, 2015). It was found that only 29 percent of respondents rated dealing with global warming as a top priority for the American president and Congress; well below the percentage that endorsed strengthening the economy (80 percent), improving the job situation (74 percent), or defending the country from terrorism (73 percent) as top priorities.

For changing human behaviour and mind set appropriate communications should be applied through the mass media. A group of psychologists (Sander van der Linden, Edward Maibach, and Anthony Leiserowitz in Tania Lombrozo, 2015) hasreviewed the psychological research to identify key aspects of climate change and climate change communication that contribute to the mismatch between the urgency and severity of climate change, on the one hand, and widespread public disengagement, on the other. They have highlighted five features of human psychology that make climate change communication especially challenging, and they pair these with recommendations for how to make science communication and policy more effective (Tania Lombrozo, 2015).

Climate change is often presented as an abstract, uncertain cost, distant in space and time, and requiring external incentives to motivate individual action. Psychological research suggests this is an especially dangerous combination, sure to make people underestimate the risk and unlikely to compel them to action. Instead, policy makers and science communicators might do well to focus on the concrete manifestations of climate change in our own experience, the consequences of warming that are affecting our communities here and now, and the ways our current actions can be tied to gains rather than losses, to social norms and to our own intrinsic motivations. So to promote the sensitivity towards climate change we have to motivate the people and encourage them in simple ways to adopt pro-environmental behaviour (Tania Lombrozo, 2015).

Through understanding emotional barriers to action, we may be able to devise better guidelines for communication, advocacy and policy. Most people who acknowledge that climate change is occurring feel that the public response has been inadequate. Psychologists have been looking at how it is that people process this information about risk and come to their understandings so that is useful to know in terms of thinking about how you can create messages that are more effective for people in terms of getting them on board (Carolyn Gregoire, 2016).

Apathy among the people towards losses due to climate change prevent them to act positively. Why do we fail to act on climate change, even when we are aware of the magnitude of the threat that lies before us? It is not just because people don't care, rather, our emotional response to the issue, which for many people is a deep but unprocessed sense of anxiety and loss, can leave us feeling powerless and paralyzed (Carolyn Gregoire, 2016). This arrested state of unprocessed grief over the destruction of the natural world, which is referred to as "environmental melancholia" blocks us from taking action. There is this feeling of loss but it hasn't been named, partly because we're not used to talking about it in our culture and society on a day-to-day basis. This is a sort of apathy among the people that though they are seeing the disaster due to climate change but are not empathize it act or they ignore it as it is not related to their own personal losses.

Changing public attitudes is one way to bring in change their behaviour towards mitigating climate change. A study was done by a group of psychologists in the UK in September 2005, by the UK Department for Transport to undertake an Evidence base review of public attitudes to climate change and travel behaviour (Anable, J; Lane, B; and Kelay, T; 2006). They found that there was only a weak link between knowledge and awareness of climate change on the one hand and travel behaviour at the individual level on the other. Raising public awareness of this link is necessary, particularly to galvanize support for carbon abatement policy, but it is not sufficient to change behaviour on its own. In order to effect change, many other factors need to be addressed. Any travel behaviour change strategy will be more effective if it targets change at the community level. Community Based Social Marketing offers a strategic framework to transform markets and behaviours.

Psychology can also be a useful way to help people feel more connected to nature. It is important to understand that how people relate to the natural world, and how to improve that relationship. The pace and structure of our modern urban lives, which are increasingly indoor, onscreen and detached from nature, have taken a toll on our relationship with the natural world. While psychologists don't yet have enough data to yet say what the exact effects might be but we have enough information to worry about (Clayton, S. & Myers, G. 2009). However, Clayton's research has pointed to a remedy. Exposing people to nature in an accessible way at their local zoos and aquariums can help to mediate feelings of disconnection from nature. She found that feelings of connection to animals at zoos or aquariums were associated with increased concern over climate change. The research also suggested that offering information about climate change at zoos helped overcome ideological barriers (Clayton, S. & Myers, G. 2009).

There is a greater need for a greater understanding of people's deep and layered psychological responses to climate change. To tackle the problem, we can't just focus on giving people information about the issue. Information is not enough, largely because people are capable of huge levels of denial. Research has shown that each narrative, by itself, fails to empower people and prevents them from being able to locate themselves and their own actions in the story about climate change. Acknowledging people's emotional responses to the crisis is a start, and offering a middle ground between facts and solutions can begin to quell anxiety and help spur action to address the problem (Clayton, S. & Myers, G. 2009).

Bringing a social norm to curb the carbon emission in society and persuade people to follow that norm. In one 2007 experiment, psychologists found that distributing flyers with the message "the majority of your neighbours are undertaking energy-saving actions every day" worked to encourage conservation better than distributing flyers emphasizing environmental impacts. In that vein, television shows that address climate change, even obliquely, could foster a norm of environmental responsibility (Nathan Collins, 2018). There is obviously no magic bullet that would alleviate the environmental crisis, but some obvious opportunities exist for constructive action and we cannot afford for the pace of social change to be slow.

Promoting pro-environmental behaviour among people is one way to promote environmental friendly behaviour. Policy-makers are interested in cost-effective and socially acceptable ways of encouraging the public to adopt more environmentally-friendly lifestyles. One area which UK policy-makers are focussing on is 'catalyst behaviour', the notion that taking-up a new behaviour (such as recycling) may cause people to adopt other pro-environmental behaviours. They assessed the influence of pro-environmental self-identify on consistency across a range of behaviours. Pro-environmental values, perceived behavioural control, subjective norm, attitudes, and demographic factors were also measured. Findings show that self-identity to be a significant behavioural determinant over and above the theory of planned behaviour variables for carbon offsetting behaviour. However, pro-environmental self-identity was only a significant predictor for certain other pro-environmental behaviours; background variables were also important predictors (Whitmarsh, Lorraine and O'Neill, Saffron, 2010).

Community participation is important for mitigating climate change such as community gardening, community cleaning of garbage and so forth. A group of researchers had introduced community gardening as a promising method of furthering well-being and resilience on multiple levels: individual, social group, and natural environment. They had examined empirical evidence for the benefits of gardening, and they had advocated the development and testing of social-ecological models of community resilience through examination of the impact of community gardens, especially in urban areas. They found the potential contribution of an extensive network of community gardens to easing the global climate change crisis and address the role of community psychologists in community gardening research and policy-oriented action (Heather A. Okvat, Alex J. Zautra, 2011).

So it could be said that through bringing the attitude change, motivation, understanding the magnitude of the risk of the climate change, promoting environmental-friendly behaviours, producing emotional states among the people, community participation etc. the human behaviour can be changed towards mitigating the climate change. Emphasis should be put on the social, cultural and technological contexts. Awareness should be brought to the public in simple languages as they do not understand the technical, scientific and demographic terminologies so that they feel about it and could be empathized the risk of climate change. Appropriate communications should be applied through the mass media. The policy makers and science communicators might do well to focus on the concrete manifestations of climate change in our own experience. Help people feel more connected to nature. Acknowledging people's emotional responses to the crisis is a start. Bring social norms to curb the carbon emission in the society and persuade people to follow. Pro-environmental self-identity should be promoted.

Cutting Consumption and Promoting Environmental-Friendly Behaviour

The most serious long-term threat facing the world is the danger that human actions are producing irreversible, harmful changes to the environmental conditions that support life on Earth. If this problem is not overcome, there may be no viable world for our descendants to inhabit. Because this threat is caused by human population growth, overconsumption, and lack of resource conservation, social scientists have a vital role in helping our world escape ecological disaster and approach a sustainable level of impact on the environmentone that can be maintained indefinitely. Enormous changes to human lifestyles and cultural practices may be required to reach this goal. We should view the achievement of sustainable living patterns as a superordinate goal- a war against the common enemy of an uninhabitable world (Stuart Oskamp, 2000).

Low carbon behaviour change is a fundamental aspect of a transition to a low carbon society. However, current government efforts to galvanize such low carbon behaviour change have been largely unsuccessful. This interdisciplinary research theme engages across the social sciences (including geography, applied policy research and social psychology) and the arts and humanities to develop better understandings of how policy could more effectively catalyze more widespread, low carbon behaviour change (David Ockwell, 2018).

Consumer policy can empower consumers for changing lifestyles by reducing personal constraints and limitations, but it should also attempt to loosen some of the external constraints that make changes towards a more sustainable lifestyle difficult. Policies that increase a feeling of empowerment may also have a positive effect on consumers' motivation to make an effort, thus amplifying its effects. Both types of constraints on lifestyle changes in a sustainable direction are necessary as well as policies for reducing constraints. Possible motivational effects of the policies are also important (Thogersen John, 2005).

One can minimise ones carbon footprint and help the environment in many different ways. Whether at home, work, school, or while you travel, small changes can add up. In brief, to reduce your carbon footprint, you'll want to do things like reduce the amount of energy you use, eat fewer animal products, shop locally, travel smart, and reduce your waste (Future Learn, 2021). Individual actions and adopting sustainable lifestyles are very important. A group of researchers examined the nature of environmental action in and around the home. Given the rise of local sustainable development and the emphasis placed on individual actions for sustainability, the role of citizens in adopting sustainable lifestyles, incorporating a range of behavioural responses from energy saving and water conservation, to waste recycling and green consumption. An approach based on a social-psychological understanding of behaviour can have significant benefits. Environmental behaviour could be seen in relation to two key issues: the way in which environmental action is framed in everyday practices (such as consumption behaviour) and how these practices are reflected amongst different segments of the population to form lifestyle groups (Stewart Barrand Andrew Gilg, 2006).

A public opinion has toemerge towards causes and consequences of climate change. Some studies revealed both a deep reservoir of concern, yet also a muddle over causes, consequences and appropriate policy measures for mitigation. The technique adopted here, namely integrated assessment focus groups, in which groups of randomly selected individuals in Switzerland looked at models of possible consequences of climate change and questioned specialists as to their accuracy and meaning, revealed a rich assembly of reactions. From the viewpoint of changing their lifestyles of material comfort and high-energy dependence, they regarded the consequences of possible behavioural shift arising from the need to meet mitigation measures as more daunting (S. Stoll Kleemann, Tim O Riordan, and Carlo C. Jaeger, 2001).

Habit formation through repeated behaviour and personal involvement is another way to bring changes in behaviour towards pro-environmental behaviour and curbing climate change. Extending existing theory in social and environmental psychology, a model was developed to study important predictors of water consumption behaviour. Overall results provide support for the predictive ability of stimuli e.g., environmental awareness, reasoned processes e.g., personal involvement, unreasoned processes e.g., habits, and situational factors e.g., income on water consumption behaviour. Findings indicate that households with lower water usage display greater awareness of water conservation issueswere more highly involved in the decision to use water, and tend to form habits associated with lower usage levels. Furthermore, the results are consistent with past research that attitudes toward water usage appear to be poor predictors of water consumption behaviour (Gary D Gregory and Michael Di Leo, 2003).

Susan A. Shaheen and Timothy E. Lipman (2007) have found that Transportation is a major contributor of carbon dioxide (CO_2) and other greenhouse gas emissions from human activity, accounting for approximately 14 percent of total anthropogenic emissions globally and about 27 percent in the U.S. Fortunately, transportation technologies and strategies are emerging that can help to meet the climate challenge. These include automotive and fuel technologies, intelligent transportation systems (ITS), and mobility management strategies that can reduce the demand for private vehicles.

Livia Albeck-Ripka (2021) a Journalist of the New York Times had suggested various ways to reduce carbon emission by bringing change in human behaviours. Real solutions will require action on a global scale, there are choices you can make in your day-to-day life to lessen your personal impact on the environment. He mentions about lessening the carbon footprints by small cautious behaviours. A carbon footprint is the total amount of greenhouse gas emissions that come from the production, use and end-of-life of a product or service. It includes carbon dioxide - the gas most commonly emitted by humans - and others, including methane, nitrous oxide, and fluorinated gases, which trap heat in the atmosphere, causing global warming. Usually, the bulk of an individual's carbon footprint will come from transportation, housing and food. No matter how you scored, here are some things that could help you lessen your personal environmental impact.

Livia Albeck-Ripka (2021) guides the reduction of carbon emission in various ways. One of the most effective ways to begin thinking about how to reduce your carbon footprint is to reconsider how much, and how often, you travel. The author suggests driving less and to try taking a train, bus or better yet, ride a bike. Make your trip more climate-friendly driving efficiently can help to reduce emissions. Regularly service your car to keep it more efficient. Check your tires. Keeping tires pumped correctly can reduce emissions. Low tire pressure will hurt your fuel economy. Shopping for a new car is a great opportunity to consider how you can reduce your personal carbon footprint. Fly often-taking one fewer long round-trips flight could shrink your personal carbon footprint significantly.

The author further suggests controlling overconsumption and the types of food one should eat which are sustainable. Cutting down on meat, and red meat in particular, is a better choice for the environment. This is because the production of red meat uses a lot of feed, water and land. Cows themselves also give off methane emissions (a harmful greenhouse gas). For that reason, eating a vegan diet is likely to be best for the environment. Eating a vegetarian diet is also likely to be better for the environment than a diet that includes a lot of meat. Overall, eating low down the food chain as often as you can is probably a good way to reduce your carbon footprint and stay healthy.

Weighing your optionswhen it comes to food, most greenhouse gas emissions happen during production, rather than transportation: What you eat is more important than where it comes from. But eating local can still make a difference. Fewer food miles can mean fewer emissions. Stop wasting the food you purchase. This is a big one: on average, Americans waste around 40 percent of the food they buy.Luckily, there are simple solutions to lower your food waste. In your home there are simple changes you can make that will save you energyand money by controlling Heat, Lights and Appliances. Make your home energy efficient with small changes to the insulation and design of your home. Buying less is the first step (Livia Albeck-Ripka, 2021).

There are simple ways to reduce the impact of your purchases. Dress sustainably which reduces carbon in manufacturing it. According to the World Resources Institute, 20 items of clothing are manufactured per person, per year. This is because of "fast fashion": clothes that are produced quickly, cheaply and unsustainably. As the price of our clothes drops, the environmental (and human) costs increase. Shop sustainably-you shop for more than just clothes, so whatever you're shopping for groceries, home goods, toys and whatever else there are ways to take the climate into account. In addition to changing your day-to-day habits, exercising your rights as a citizen is one of the most significant things you can do to help the planet. Taking climate change into consideration when you vote is a good start. Recycling waste is another way to save the environment. American's recycled and composted millions of tons of municipal solid waste which saved the same amount of energy as generated by 25 million homes (Livia Albeck-Ripka, 2021).

More are less the similar guidelines have been suggested by Renee Cho (2018), and by Kelly Oakes (2020). However, Diana Ivanova (2020) suggests that our society needs to drastically reduce carbon footprints to limit climate change. This requires tightening environmental standards and policies, greening infrastructure and service provisioning, and re-thinking consumption. Crucially, these climate considerations should govern policy and action worldwide, including the response to the COVID pandemic. Our planet cannot sustain this overconsumption. We need ambitious policy and industry climate responses, which make binding commitments to the climate targets and act by their commitments. Carbon footprints are also very unequal across the globe, with the wealthy usually being the biggest polluters. Placing the focus on fair distribution is the key in the aim to satisfy human needs within planetary boundaries.

So cutting consumption by using less and minimum required resources is the key to reducing carbon emission and other gases. So we have to check the human population growth, overconsumption, and lack of resource conservation. Enormous changes to human lifestyles and cultural practices may be required to reach this goal. The current government should make efforts to galvanize such low carbon behaviour change. Proper consumer policy can be brought for changing lifestyles by reducing personal constraints and limitations towards a more sustainable lifestyle. Carbon footprints could be minimized inmany different ways whether at home, work, school, or while you travel. A public opinion has to emerge towards causes and consequences of climate change. Habit formation through repeated behaviour should be promoted to bring changes towards pro-environmental behaviour and curbing climate change. Traveling by private cars is another major source of carbon emission. This should be curbed using public transport or bike. An action on a global scale is required. One can make it in one's day-to-day life to lessen the personal impact on the environment.

Conclusions

Human being on earth is the career of the climate change. His behaviour is responsible for the factors induced by environmental degradation and global warming. So by bringing positive changes in human behaviour climate change could be mitigated. Changing behaviour, changing lifestyles and promoting pro-environmental behaviour is not easy. A group of psychologists have worked towards this end and have suggested various measures to curb carbon emission and stop environmental degradation. Many studies have resulted in changing attitudes, mind sets, perceptions, group involvements, community participation, and awareness about the dangers of climate change in the future. Bringing the real pictures in peoples' mind through providing information and knowledge to change their cognition about climate change and motivate them to take collective responsibility to control it. Certain guidelines have been suggested by some psychologists to stop overconsumption of food, home appliances, manufactured goods and any other natural resources. One should not to sway with the consumerist market forces but be within the limit of the real needs. Garbage managements, recycling of waste materials, plantation and gardening, minimizing the use of personal cars and using public transport or bikes are the possible ways to promote environment-friendly

behaviours. The government of the world should bring various policies to stop climate change by proving positive incentives to the positive human behaviours to mitigate climate change.

References:

Anable, J; Lane, B; and Kelay, T. (2006). An Evidence Base Review of Public Attitudes To Climate Change and Transport Behaviour. Final Report, for the Department For Transport July 2006, UK.

APA Task Force Report (2011). Psychology and Global Climate Change: Addressing a Multi-faceted Phenomenon and Set of Challenges. Report of the American Psychological Association Task Force on the Interface Between Psychology and Global Climate Change. American Psychologist Special Issue (May-June 2011). Retrieved on August 2018 from http://www.apa.org/science/about/publications/climate-change.aspx

Carolyn Gregoire (2016). Why Psychology Should Be A Part of The Fight Against Climate Change: By understanding emotional barriers to action, we may be able to devise better guidelines for communication, advocacy and policy. SCIENCE, The Huffington Post, 01/18/2016 11:40 am ET. Retrieved on August 19, 2021 from https://www.huffpost.com/entry/climate-change-psychology

Clayton, S. & Myers, G. (2009). Conservation Psychology: Understanding and Promoting Human Care for Nature (Wiley-Blackwell, 2009).

Clayton, S., & Brook, A. (2005). Can psychology help save the world? A model for conservation psychology. Analyses of Social Issues and Public Policy (ASAP), 5(1), 87-102.

Clayton, S., and Brook, A. (2005). Can psychology help save the world? A model for conservation psychology. Analyses of Social Issues and Public Policy (ASAP), 5(1), 87-102.

Cvetkovich, G.T. and Wener, R. (1994). How can psychology help save the planet: A research agenda on environmental problems. Statement distributed by the APA Taskforce on Psychology and Environmental Problems. Washington, D.C.:

David Ockwell (2018). Low carbon behaviour change. Department of Geography, University of Sussex Retrieved on Augus 14, 2018 from http://www.sussex.ac.uk/geography/research/climateresearch/low_carbon_behaviour

Diana Ivanova (2020). Challenging consumption to cut carbon foot prints.

CREDS. Retrieved on August 2021 from https://www.creds.ac.uk/challenging-consumption-to-cut-carbon-footprints/

Fischhoff, B., & Furby, L. (1983). Psychological dimensions of climatic change. In Social Science research and climate change: An interdisciplinary appraisal. R. S. Chen, E. Boulding, S. H. Schneider (eds), pp. 180-203. Dordrecht: Reidel. Retrieved on August 14, 2018 from http://www.apa.org/about/policy/climate-change.aspx

Fischhoff, B., and Furby, L., (1983). Psychological dimensions of climatic change. In Social Science research and climate change: An interdisciplinary appraisal. R. S. Chen, E. Boulding, S. H. Schneider (eds), pp. 180-203. Dordrecht: Reidel.

Future Learn (2021). How to reduce your carbon footprint - 20 top tips. Published on January 12th, 2021 by Future Learn. Retrieved on August 19, 2021 from https://www.futurelearn.com/info/blog/how-to-reduce-your-carbon-footprint-tips

Gary D Gregory and Michael Di Leo (2003). Repeated Behavior and Environmental Psychology: The Role of Personal Involvement and Habit Formation in Explaining Water Consumption. Journal of Applied Social Psychology, DOI: 10.1111/j.1559-1816.2003.tb01949.xView/save citation. Retrieved on a August 18, 2018 from http://onlinelibrary.wiley.com/doi/10.1111/j.1559-1816.2003.tb01949.x/full

Gifford, R. (2007). Environmental Psychology and Sustainable Development: Expansion, Maturation, and Challenges. Journal of Social Issues. 63(1), 199-212.

Gifford, R. (2007). Environmental Psychology and Sustainable Development: Expansion, Maturation, and Challenges. Journal of Social Issues. 63(1), 199-212. Retrieved on August 14, 2018 from http://www.apa.org/about/policy/climate-change.aspx

Heather A. Okvat, Alex J. Zautra (2011). Community Gardening: A Parsimonious Path to Individual, Community, and Environmental Resilience. Americal Journal of Community Psychology, Volume47, Issue3-4, June 2011, Pages 374-387. https://onlinelibrary.wiley.com/doi/full/10.1007/s10464-010-9404-z

International Disability and Development Consortium (2008). Statement to Human Rights Council Resolution 7/23 "Human Rights and Climate Change. Retrieved on August 14, 2018 from http://www.apa.org/about/policy/climate-change.aspx

Kazdin, A. E. (2009). Psychological science's contributions to a sustainable environment:

Extending our reach to a grand challenge of society. American Psychologist. 64(5), 339-356.

Kelly Oakes (2020). In a guideline How cutting your food waste can help the climate. Retrieved on August 19, 2021 from https://www.bbc.com/future/article/20200224-how-cutting-your-food-waste-can-help-the-climate

Kelly Oakes (2020). In a guideline How cutting your food waste can help the climate. 26th February, 2020. Retrieved on August 14, 2018 from http://www.apa.org/about/policy/climate-change.aspx

Livia Albeck- Ripka (2021). How to reduce your carbon foot prints. Newyork Times: A year of Leaving Better Retrieved on August 19, 2021 from https://www.nytimes.com/guides/year-of-living-better/how-to-reduce-your-carbon-footprint

Lombrozo Tania (2015). How Psychology Can Save the World from Climate Change. November 30, 201510:52 AM ET. Retrieved on August 14, 2018 from http://www.npr.org/sections/13.7/2015/11/30/457835780/how-psychology-can-save-the-world-from-climate-change

Moore Jeanne (2012). Social and Behavioural Aspects of Climate Change Background Paper No. 3. Final Report from the NESC Secretariat: Ireland's Climate Change Challenge: Connecting 'How Much' with 'How To'. National Economic and Social Council, Ireland. December 2012.

Nathan Collins (2018). The psychology of climate change inaction.Pacific Standard.http://theweek.com/articles/619768/psychology-climate-change-inaction

Nathan Collins (2018). The psychology of climate change inaction. Pacific Standard. http://theweek.com/articles/619768/psychology-climate-change-inaction

National Research Council (2010). Advancing the science of climate change. Washington: National Academies Press. Retrieved on August 14, 2018 from http://www.apa.org/about/policy/climate-change.aspx

Renee Cho (2018). By Suggesting 35 Easiest Ways to Reduce Carbon Foot Prints. December 27, 2018. Retrieved on August 14, 2018 from http://www.apa.org/about/policy/climate-change.aspx

Renee Cho (2018). The 35 easiest ways to your carbon foot prints. From the series of sustainable living. State of the planet, news from the Columbia climate school. Retrieved on august 19, 2021 from https://news.climate.columbia.edu/2018/12/27/35-ways-reduce-carbon-footprint/

S. Stoll-Kleemann, Tim O'Riordan, and Carlo C. Jaeger (2001). The psychology of denial concerning climate mitigation measures: evidence from Swiss focus groups. Global Environmental Change, Volume 11, Issue 2, July 2001, Pages 107-117. Retrieved on August 14, 2018 from http://dx.doi.org/10.1016/S0959-3780(00)00061-3

Stewart Barrand Andrew Gilg (2006). Sustainable lifestyles: Framing environmental action in and around the home. Geoforum, Vol 37, Issue 6, November 2006, Pages 906-920. http://dx.doi.org/10.1016/j.geoforum.2006.05.002

Stuart Oskamp (2000). Psychology of Promoting Environmentalism: Psychological Contributions to Achieving an Ecologically Sustainable Future for Humanity. Journal of Social Issues, Volume 56, Issue 3, Fall 2000, Pages 373-390. https://doi.org/10.1111/0022-4537.00173 Retrieved on August 18, 2018 from http://onlinelibrary.wiley.com/doi/10.1111/0022-4537.00173/full

Susan A. Shaheen and Timothy E. Lipman (2007). Reducing Greenhouse Emissions and Fuel Consumption: Sustainable Approaches for Surface Transportation. IATSS Research, Volume 31, Issue 1, 2007, Pages 6-20. https://doi.org/10.1016/S0386-1112(14)60179-5 Retrieved on August 19, 2021 from https://www.sciencedirect.com/science/article/pii/S0386111214601795

Tania Lombrozo (2015). How Psychology Can Save The World From Climate Change. Commentary, November 30, 201510:52 AM ET. Retrieved on August 14, 2018 from http://www.npr.org/sections/13.7/2015/11/30/457835780/how-psychology-can-save-the-world-from-climate-change

Thogersen John (2005). How May Consumer Policy Empower Consumers for Sustainable Lifestyles? Journal of Consumer Policy, June 2005, Volume 28, Issue 2, pp 143-177. DOI: 10.1007/s10603-005-2982-8.

Uzzell, D., & Moser, G. (2009). Introduction: Environmental psychology on the move. Journal of Environmental Psychology. 29(3), 307-308. Retrieved on August 14, 2018 from http://www.apa.org/about/policy/

climate-change.aspx

Uzzell, D., and Moser, G. (2009). Introduction: Environmental psychology on the move. Journal of Environmental Psychology. 29(3), 307-308. http://www.apa.org/about/policy/climate-change.aspx

Van der Werff, E., Steg, L. & Keizer, K. (2013). The value of environmental self-identity: The relationship between biospheric values, environmental self-identity and environmental preferences, intentions and behaviour. J. Environ. Psychol. 34, 55-63.

Whitmarsh, Lorraine and O'Neill, Saffron (2010). Green identity, green living? The role of pro-environmental self-identity in determining consistency across diverse pro-environmental behaviours. Journal of Environmental Psychology, Volume 30, Issue 3, September 2010, Pages 305-314. http://dx.doi.org/10.1016/j.jenvp.2010.01.003. Retrieved on August 19, 2021 from http://www.sciencedirect.com/science/article/pii/S0272494410000046